GIANT BOOK OF

Science Experiments

H.J. Press

Sterling Publishing Co., Inc.
New York

Library of Congress Cataloging-in-Publication Data

Press, Hans Jürgen
    Giant book of science experiments / [Hans Jürgen Press]
      p. cm.
    "Originally jpubublished in German by Ravensburger Buchverlag
Otto Maier GmbH under the titles: Spiel, das Wissen schafft; Der Natur
auf der Spur; Geheimnisse des Altags"—Verso t.p.
    Include index.
    ISBN 0-8069-8139-3
    1. Science—Experiments   I. Title
  Q182.3.p75   1998
  507'.8—dc21   97–43003
                                             CIP

10 9 8 7 6 5 4 3 2 1

Published by Sterling Publishing Company, Inc.
387 Park Avenue South, New York, N.Y. 10016
The projects in this compilation originally appeared in German in the fol-
    lowing volumes published and copyright by Ravensburger Buchverlag
    Otto Maier GmbH: Spiel—das Wissen schafft © 1964, 1995: Der Natur
    auf der Spur © 1972, 1995; Geheimnisse des Altags © 1977, 1995.
One volume edition published in Germany © 1995 by Buchverlag Otto
    Maier GmbH
English Translation © 1998 by Sterling Publishing Co., Inc.
Distributed in Canada by Sterling Publishing
c/o Canadian Manda Group, One Atlantic Avenue, Suite 105
Toronto, Ontario, Canada M6K 3E7
Distributed in Great Britain and Europe by Cassell PLC
Wellington House, 125 Strand, London WC2R 0BB, England
Distributed in Australia by Capricorn Link (Australia) Pty Ltd.
P.O. Box 6651, Baulkham Hills, Business Centre, NSW 2153, Australia

Sterling ISBN 0-8069-8139-3

Contents

Inertia

Sound

Light

Illusions

CONTENTS

Birds

Reptiles and amphibians

Fish

Invertebrates

Miscellaneous nature

Introduction

With over three hundred experiments and natural observations that are both entertaining and educational, this is the perfect book for the young person who wants to discover how things works and why things are the way they are. As the experimenter learns to use a wristwatch as a compass, light a fire with ice, or create rain indoors, he or she learns the underlying scientific principles. Each project gently introduces the reader to some fascinating aspect of subjects ranging from electricity, magnetism, heat, and light to botany, birds, and invertebrates. Each project is illustrated and includes clear instructions and explanations of the cause and effect.

With appropriate adult supervision, each project is safe and easy to do. The only materials needed are simple tools and ordinary household articles. Children of all ages, as well as their parents, will find that this book both stimulates their mind and satisfies their curiosity.

1. Small sun pictures

When the sun is standing high in the sky, circular spots of light show up on the ground in the shadow of large trees. Why is it that they are not shaped irregularly, like the gaps between the leaves? The sunlight, which falls onto the ground through the gaps in the foliage, projects small images of the sun. The smaller the gap, the sharper is the image. Each gap works like the stop of a camera: it keeps out interfering light from the sides and lets only slender rays of light come through, which create a sharp image.

During a solar eclipse, when the moon partially covers the sun, the little images of the sun in the tree shadows also change their form. They appear as little crescents.

2. Image of the sun

Looking at the sun directly, or through binoculars, can cause severe eye damage or blindness. But here is a safe way to observe the sun. Place a pair of binoculars in an open window in the direct path of the sun's rays. Stand a mirror in front of one eyepiece so that it throws an image of the sun onto the opposite wall of the room. Adjust the mirror until the image is sharp, and darken the room.

You can view the bright disk on the wall as large and clear as in the movies. Clouds and birds passing over can also be distinguished and, if the binoculars are good, even sunspots. These are cooler areas on the glowing sphere, some so big that many earths could fit into them. Because of the earth's rotation, the sun's image moves quite quickly across the wall. Do not forget to re-align the binoculars from time to time onto the sun. The moon and stars cannot be observed in this way because the light coming from them is too weak.

3. Sundial

Place a flowerpot with a long stick fixed into the hole at the bottom in a spot which is sunny all day. The stick's shadow moves along the rim of the pot as the sun moves. Each hour mark the position of the shadow on the edge of the pot. Thereafter, if the sun is shining, you can read off the time.

Because of the rotation of the earth, the sun appears to pass over us in a semicircle. In the morning and evening its shadow strikes the pot at a low angle, while at midday, around 12 o'clock, the light shines down at its steepest angle to the ground. The shadow can be seen particularly clearly on the sloping wall of the pot.

4. A watch as a compass

Hold a watch horizontally, with the hour hand pointing directly to the sun. If you place a match halfway between the hour hand and the 12, the end of the match points directly to the south.

In 24 hours the sun appears to move, because of the earth's rotation, once around the earth. But the hour hand of the watch goes twice around the dial. Therefore, before noon we halve the distance from the hour hand to the 12, and after midday from the 12 to the hour hand. The match always points to the south. At midday, or 12 o'clock, the hour hand and the 12 both point to the sun standing in the south.

5. Color magic

Cut a red cabbage leaf into small pieces and soak them in a cup of boiling water. After half an hour pour the violet-colored cabbage water into a glass. You can now use it for crazy color magic. Place three glasses on the table, all apparently containing pure water. In fact only the first glass contains water; in the second is white vinegar and in the third water mixed with bicarbonate of soda (baking soda). When you pour a little cabbage water into each glass, the first liquid remains violet, the second turns red, and the third green. The violet cabbage dye has the property of turning red in acid liquids and green in alkaline. In neutral water it does not change color. In chemistry you can find out whether a liquid is acid or alkaline by using similar detecting liquids ("pH" indicators).

6. Violet becomes red

If you ever come across an anthill, you can conduct a small chemical experiment. Hold a violet flower, such as a bluebell, firmly over the ants. The insects feel threatened and spray a sharp-smelling liquid over the flower. Where the flower is hit it turns red.

The ants make a corrosive protective liquid in their hind quarters, called *formic acid*. You notice it if an ant nips you, though it is generally quite harmless. The acid turns the blue pigments in the flower red.

7. Invisible ink

If you ever want to write a secret message on paper, simply use vinegar, lemon, or onion juice as the invisible ink. Write with one of these on white writing paper. After it dries the writing is invisible. The person who receives the letter must know that the paper has to be held briefly over a candle flame; the writing turns brown and is clearly visible. Be careful to not let the paper catch fire.

Vinegar and lemon or onion juice cause a chemical change in the paper, converting it to a substance similar to cellophane. Because its ignition temperature is lower than that of the paper, the parts you have written on singe and turn brown.

8. Bleached rose

A piece of sulfur is ignited in a jam jar. Since a pungent vapor is produced, you should do the experiment outside. Hold a red rose in the jar. The color of the flower becomes visibly paler until it is white.

When sulfur is burned, sulfur dioxide is formed. The gas has a bleaching effect, which destroys the dye in the flower. Sulfur dioxide also destroys the chlorophyll of plants, which explains their poor growth in industrial areas, where this gas pollutes the air.

9. Sugar fire

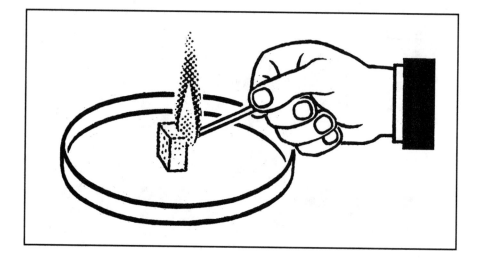

Place a sugar cube on a tin lid and try to set it on fire. You will not succeed. However, if you dab a corner of the cube with a trace of ash from a burned piece of paper and hold a burning match there, the sugar begins to burn with a blue flame until it is completely gone.

The ash and sugar cannot be separately ignited, but together the ash initiates the combustion of the sugar. We call a substance which brings about a chemical reaction, without itself being changed, a catalyst.

10. Jet of flame

Light a candle, let it burn for a while, and blow it out. White smoke rises from the wick. If you hold a burning match in the smoke close to the candle, a jet of flame shoots down to the wick, and it relights.

After the flame is blown out, the stearin (a chemical in the wax) is still so hot that it continues to evaporate and produce a vapor. Because it is combustible, it can be easily relighted by a naked flame. The experiment shows that solid substances must first turn into a gas before they will burn in a supply of oxygen.

11. Gas pipe

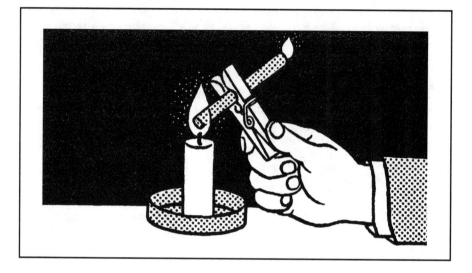

Roll a thin piece of aluminum foil around a pencil and remove it to make a tube about four inches long. Hold it with one end in the middle of a candle flame. If you hold a burning match at the other end of the tube, a second flame will be lit there.

Like all solid and liquid fuels, stearin produces combustible gases when heated, and these accumulate inside a flame. They burn, with the oxygen of the air, in the outer layer and tip of the flame. The unburned stearin vapor in the middle can be drawn off and ignited.

GIANT BOOK OF SCIENCE EXPERIMENTS

12. Bubbles from the mud

You can see bubbles rise from the muddy bottom of a pond, especially during the summer. If you poke a stick into the bottom, the bubbles gush up to the surface. What are they?

The bubbles are not exhaled air from aquatic animals, but rather methane gas, which is created during the process of fermentation and decomposition of plant remains in the mud. For a hint, try to light the bubbles with a match; they will explode with a blue flame.

Methane gas, a hydrocarbon, is flammable. It is the predominant component of the natural gas used for heating and cooking, which has formed over millions of years as organic material buried in the upper layers of the earth decayed.

CHEMISTRY

13. Gas balance

Fix two plastic bags to the ends of a narrow piece of wood about 18 inches long and let it swing like a balance on a thumbtack. Pour some bicarbonate of soda (baking soda) and some vinegar into a glass. It begins to froth, because a gas is escaping. If you tilt the glass over one of the bags, that side of the balance falls.

The gas which is given off during the chemical reaction is carbon dioxide. It is heavier than air, so it can be poured into the bag to add weight to it. If you were to fill a balloon with this gas it would never rise. For that purpose, other gases that are lighter than air are used.

14. Fire extinguisher

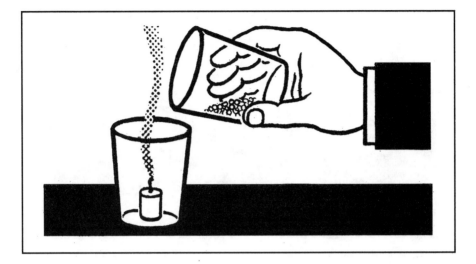

Light a candle stump in an empty glass, and mix in another glass a teaspoonful of bicarbonate of soda (baking soda) with some vinegar and let it froth. If you tilt the glass over the candle, the flame goes out.

Because the carbon dioxide formed in the chemical reaction in the top glass is heavier than air, it displaces the oxygen needed for the flame. The noncombustible carbon dioxide smothers the flame. Many fire extinguishers work in the same way: the sprayed foam consists of bubbles filled with carbon dioxide. It surrounds the flame and blocks the supply of oxygen.

15. Burning without a flame

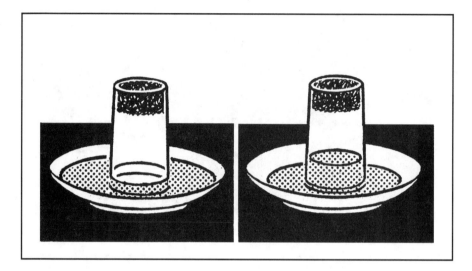

Press a handful of steel wool firmly into a glass and moisten it. Invert the glass over a dish containing water. At first the air in the glass prevents the water from entering, but soon the level of water in the dish becomes lower while that in the glass rises.

After the steel wool is moistened, it begins to rust. The iron combines with the oxygen in the air—a process we call combustion, or oxidation. An imperceptible amount of heat is set free in the process. Since the air consists of about one-fifth oxygen, the water rises in the glass until after several hours it fills one-fifth of the space.

16. Burning iron

Would you have thought that even iron could be made to burn with a flame? Twist some fine steel wool around a small piece of wood and hold it in a candle flame. The metal begins to blaze and scatter sparks like a sparkler.

The oxidation, which was slow in the previous experiment, is rapid in this case. The iron combines with the oxygen in the air to form iron oxide. The temperature this produces is higher than the melting point of iron. Because of the falling red-hot particles of iron, it is advisable to carry out the experiment in a basin.

17. Destroyed metal

Put a piece of aluminum foil with a copper coin on it into a glass of water, and let it stand for a day. After this period, the water looks cloudy and where the coin was lying the aluminum foil is perforated.

This process of decomposition is known as corrosion. It often occurs at the point where two different metals are directly joined together. With metal mixtures (alloys) it is particularly common if the metals are not evenly distributed. In our experiment the water becomes cloudy due to dissolved aluminum. A fairly small electric current is also produced in this process.

18. Flowing electrons in a battery

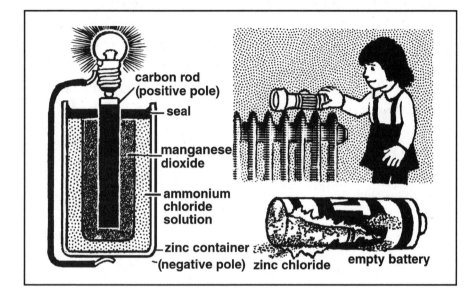

carbon rod
(positive pole)

seal

manganese
dioxide

ammonium
chloride
solution

zinc container
(negative pole)   zinc chloride   **empty battery**

When a flashlight no longer works, it does not necessarily mean that the battery is used up. Try warming the battery up; the bulb will usually light up again. Why is that?

In a chemical reaction, the zinc container of a battery cell is corroded by the ammonium chloride solution chemically, which is in the form of a paste. This creates the potential for a flow of electricity by producing an excess of electrons in the zinc and an electron deficiency in the carbon rod (which is coated with manganese dioxide to prevent the buildup of hydrogen gas that would otherwise stop the reaction). When you connect a bulb to the battery, electrons flow away from the zinc (negative pole) to the carbon rod (positive pole). The bulb's incandescent filament lights up, as long as sufficient electrons flow through it.

When the battery gets old, the chemical reaction slows down and the flow of the electrons gets too weak to make the filament glow anymore. Heating up the battery accelerates the chemical reaction so that it can briefly light the bulb again. When the zinc has corroded entirely and turned into white powder, zinc chloride, it is finished.

19. Coin current

Place several copper pennies (use the older ones, which are solid copper) and pieces of sheet zinc of the same size alternately above one another, and between each metal pair insert a piece of blotting paper soaked in saltwater. Electricity, which you can detect, is produced. Wind thin, insulated copper wire about 50 times around a compass, and hold the bare ends on the uppermost coin and on the last zinc disk. The current causes a deflection of the compass needle.

In a similar experiment the Italian physicist Volta generated a current. The salt solution acts on the metal.

20. Graphite conductor

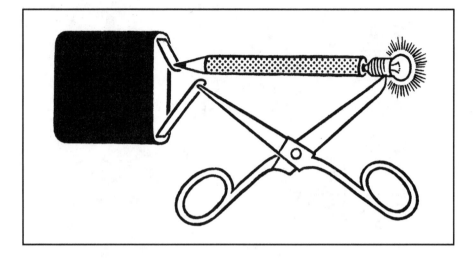

Connect a flashlight bulb with a battery by means of a pair of scissors and a pencil. The bulb lights up.

From the negative pole of the battery the current (of electrons) flows through the metal of the scissors to the bulb. Because the bulb wire resists the flow of electricity, the bulb filament heats up and glows. The current then flows through the graphite shaft to the positive pole of the battery. Graphite in the "lead" pencil is a good conductor; so much electricity flows even through graphite on paper that you can hear crackling in earphones.

21. Microphone

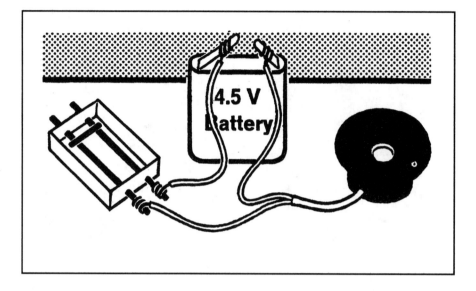

Push two pencils leads through the short sides of a matchbox, just above the base. Scrape off some of the surface, and do the same with a shorter lead, which you lay across the top. Connect the microphone with a battery and earphone in the next room. (You can take the earphone from a transistor radio.) Hold the box horizontally and speak into it. Your words can be heard clearly in the earphone.

The current flows through the graphite "leads." When you speak into the box, the base vibrates, causing pressure between the "leads" to alter and making the current flow unevenly. The current variations cause vibrations in the earphone.

22. Mysterious circles

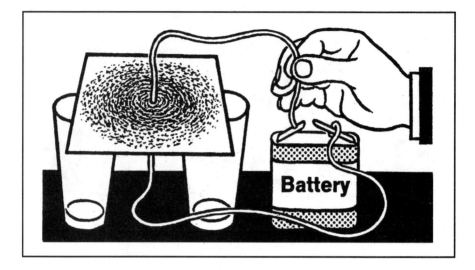

Push a length of copper wire through a piece of cardboard laid horizontally, and connect the ends of the wire to a battery. Scatter iron filings onto the cardboard and tap it lightly with your finger. The iron filings form circles around the wire.

If a direct current is passed through a wire or another conductor, a magnetic field is produced around it. The experiment would not work with an alternating current, in which the direction of the current changes in rapid sequence, because the magnetic field would also be changing continuously.

GIANT BOOK OF SCIENCE EXPERIMENTS

23. Electromagnet

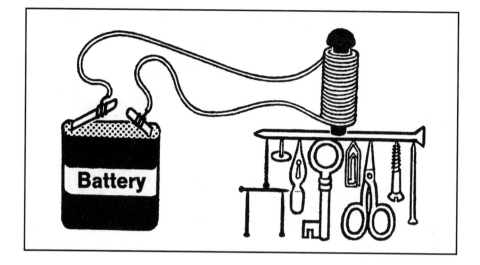

Wind one to two yards of thin insulated wire around an iron bolt and connect the bare ends of the wire to a battery. The bolt will attract all sorts of metal objects.

The current in the coil produces a magnetic field. The tiny magnet particles in the iron become arranged in an orderly manner, so that the iron has a magnetic north and south pole. If the bolt is made of soft iron, it loses its magnetism when the current is switched off, but if it is made of steel it retains it.

24. Electric buzzer

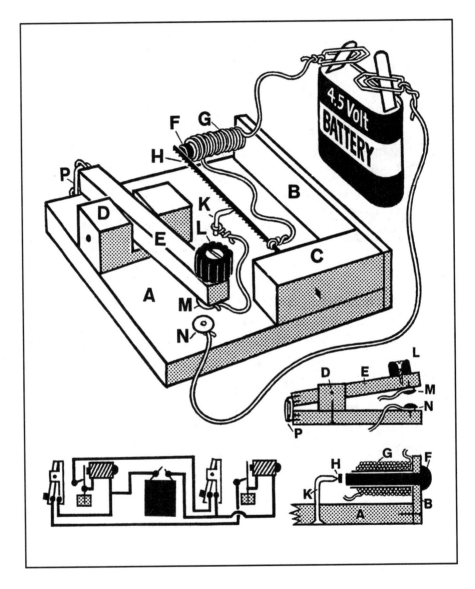

Nail board B and wooden blocks C and D onto board A (about 5 × 5 inches). Push an iron bolt F through a hole bored in B. Wind insulated copper wire G 100 times around the bolt and connect one end to a battery. Bore a hole through block C and wedge the coping saw blade H firmly into it so that its end is a short distance from bolt F. Attach the other end of the wire to the blade. Hammer a long nail K through A and bend it so that its point rests in the middle of the saw blade. Oil the point of the nail. Use a piece of wood E as a key, with a rubber band P as spring and thumbtacks M and N as contacts. Join all the parts with connecting wire (remove the insulation).

If you press the key down, you connect the electric circuit and bolt F becomes magnetic and attracts H. At this moment the circuit is broken at K and the bolt loses its magnetism. H jumps back and reconnects the current. This process is repeated so quickly that the saw blade vibrates and produces a loud buzz. If you wish to do Morse signaling with two pieces of apparatus, you must use the setup shown in the lower circuit diagram.

25. Potato battery

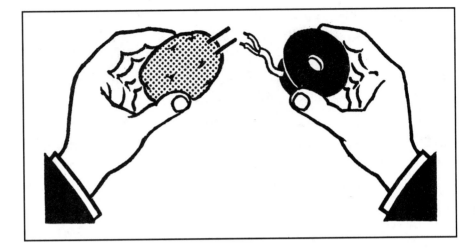

Stick finger-length pieces of copper and zinc wire one at a time into a raw potato. If you hold an earphone on the wires, you will hear a distinct crackling.

The noise is caused by an electric current. The potato and wires produce an electric current in the same way as a flashlight battery, but only a very weak one. The sap of the potato reacts with the metals in a chemical reaction that produces electrical energy. We call this a type of a galvanic cell because the Italian doctor Galvani first observed this process in a similar experiment in 1789.

26. Light fan

Hold a light-colored rod between your thumb and forefinger and move it quickly up and down in neon light. You do not see, as you might expect, a single blurred, bright line, but what appears to resemble an oriental fan with light and dark ribs.

Neon tubes contain a gas, which flashes on and off 60 (in U.S.) times a second because of rapid reversals in alternating current. The moving rod is thrown alternatively into light and darkness in rapid sequence, so that it seems to move by jerks in a semicircle. The light from a television set will produce the same effect. Normally, the eye is too slow to notice these breaks in illumination clearly. In a regular electric light bulb, the metal filament continues glowing between the peaks in current.

27. Dangerous step tension

If you are caught in a thunderstorm, you are not supposed to run. Lightning, which hits a tree nearby, can endanger a runner as its electrical energy flows through the ground. Since the potential energy decreases with increasing distance from the tree, it can be stronger under the rear leg of the runner than under the front one. Because electricity flows towards the lowest potential, this "step tension" causes the current to pass through the body, since it conducts the electricity better than the ground.

A person is safer squatting with his feet together in a depression in the ground to avoid bolts from above. The electricity flows away underneath him, as if he were a bird on a power line.

GIANT BOOK OF SCIENCE EXPERIMENTS

28. Path of the electrical current

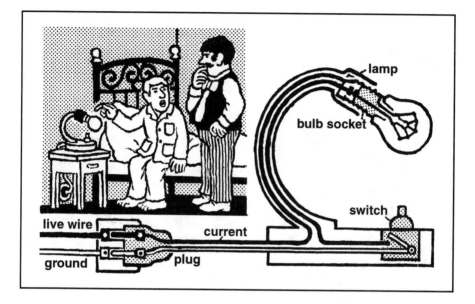

A hotel guest complains about a night lamp which is not up to standard. When he was searching for the switch in the dark, he received an electric shock from the bulb socket, which juts a little bit out of the lamp. The proprietor doubts that he was shocked, arguing that the lamp had been switched off, thus preventing any electricity from flowing. Who is right?

The answer depends on how the plug sits in the outlet. If the wire that conducts the alternating current leads directly to the switch, the current is interrupted there. But if the current flows first to the bulb, via the other wire, then the lamp's socket, the bulb's glowing wire, and the rest of the circuit leading to the switch are under current, without the bulb being on, creating a shock hazard.

ELECTRICITY

29. Bicycle circuit

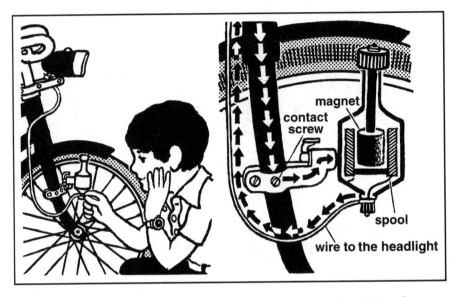

magnet

contact
screw

spool

wire to the headlight

A boy wonders why only one wire leads from his bicycle generator to the headlight; there should be two wires to complete a closed circuit. So how does the electricity flow from the generator to the headlight?

When the bicycle is in motion, a permanent magnet inside the generator rotates in the center of a spool of tightly wound copper wire. The magnetic force of the spinning magnet creates an electrical potential, or tension, in the copper windings. Electricity flows through the wire to the headlight, through the glowing filament of the bulb, and then through the headlight casing, the bicycle fork, and the metal casing of the generator, back to the spool. The small contact screw in the generator casing is important; it pierces through the insulating layer of paint to the metal of the fork and closes the circuit.

30. Clinging balloons

Blow up some balloons, and rub them for a short time on a wool sweater. If you place them against the ceiling, they will remain there for hours.

The balloons become negatively charged with static electricity when they are rubbed; that is, they remove minute, negatively charged particles, called electrons, from the sweater. Because electrically charged bodies are weakly attracted to those which are neutral, or uncharged, the balloons cling to the ceiling until the charges between the two gradually become equal. This generally takes hours in a dry atmosphere, because the electrons flow slowly from the balloons into the ceiling, which is a poor conductor.

31. Pepper and salt

Scatter some coarse salt onto the table and mix it with some ground pepper. How are you going to separate them again? Rub a plastic spoon (or several for the best effect) with a wool cloth and hold it about an inch or so above the mixture. The pepper jumps up to the spoon and sticks to it.

The plastic spoon becomes negatively charged when it is rubbed, causing it to attract both the pepper and salt. The pepper rises first, however, because it is lighter than the salt. To catch the salt grains, you must hold the spoon lower, increasing the force of the attraction.

32. Coiled snake

Cut a spiral-shaped coil from a thin piece of tissue paper about 4 inches square, lay it on a tin lid, and bend its head up. Rub a fountain pen vigorously with a wool cloth and hold it over the coil. It rises like a living snake and reaches upwards.

In this case the fountain pen has taken electrons from the woollen cloth and attracts the uncharged paper. On contact, the paper falls because it takes part of the negative electric charge and gives it up immediately to the metal lid, which is a good conductor. Since the paper is now uncharged again, it is again attracted upwards until the fountain pen has lost its charge.

33. Water bow

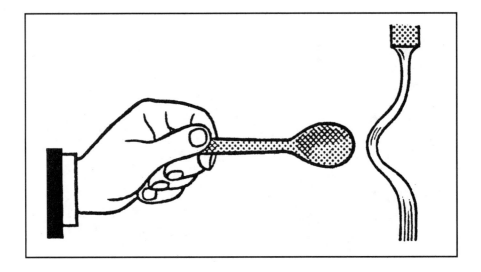

Once more rub a plastic spoon with a woollen cloth. Turn a water tap on gently and hold the spoon near the fine stream. The water will be pulled towards the spoon in a bow.

The electric charge attracts the uncharged water particles. If the water touches the spoon, however, the effect is lost. Water conducts electricity and draws the charge from the spoon. Tiny water particles suspended in the air also take up electricity. Therefore, experiments with static electricity always work best on clear days and in centrally heated rooms.

34. Hostile balloons

Blow up two balloons and join them with string. Rub both on a wool sweater and let them hand downwards from the string. Instead of being attracted like the objects in the previous experiments, they float away from each other.

When rubbed, both balloons become negatively charged because they have taken electrons from the sweater, which has now gained a positive charge. Negative and positive charges attract each other, so the balloons will stick to the sweater. Similar charges, however, repel one another, so the balloons move away from each other.

35. Simple electroscope

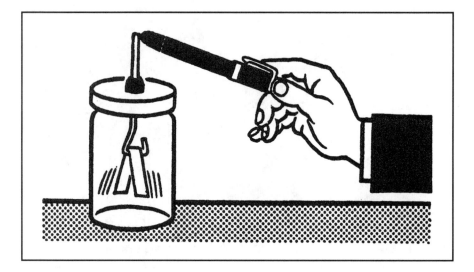

Bore a hole through the lid of a jam jar and push a piece of copper wire bent into a hook through the hole. Hang a folded strip of aluminum foil over the hook. If you hold a fountain pen, comb, or similar object which has been electrically charged by rubbing on the top of the wire, the ends of the strip spring apart.

On contact with a charged object, electrons flow through the wire to the ends of the strip. Both now have the same negative charge and repel one another according to the strength of the charge, which is what an electroscope measures.

36. Electrical ball game

Tape a piece of aluminum foil cut into the shape of a soccer player on the edge of a phonograph record that you no longer play. Rub the record vigorously with a wool cloth and place it on a dry glass. Put a tin can about two inches in front of the figure. If you hold a small aluminum-foil ball on a thread between them, it swings repeatedly from the figure to the can and back.

The negative electric charge on the record flows into the aluminum-foil figure and attracts the ball. When they touch, the ball becomes negatively charged, but is immediately repelled because the charges become equal, and it goes to the can, where it loses its electrons. This process is repeated for a time.

37. Electric fleas

Rub an old phonograph record with a wool cloth and place it on a glass. If you toss some small aluminum-foil balls onto the record, they will jump away from one another in a zigzag motion. If you then move the balls together with your fingers, they will hop fiercely away again.

The electricity produced on the record by rubbing is distributed in irregular fields. The balls take up the negative charge and are repelled, but they are again attracted to fields with the opposite (positive) charge. They will also be repelled when they meet balls with the same charge.

38. Puppet dance

Lay a pane of glass across two books, with a metal plate underneath. Cut out tiny figures an inch or so high from tissue paper. If you rub the glass with a woollen cloth, the figures underneath begin a lively dance. They stand up, turn around in a circle, fall, and spring up again.

The glass becomes electrically charged when it is rubbed with the wool, attracts the dancers, and also charges them. Since the two like charges repel each other, the figures fall back on the plate, give up their charge to the metal, and are again attracted to the glass.

39. Cleaning the air

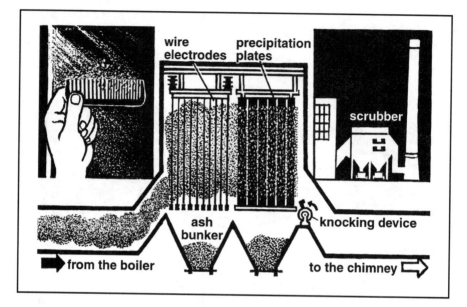

wire electrodes · precipitation plates · scrubber · ash bunker · knocking device · from the boiler · to the chimney

Look at a sunbeam passing through a crack into a darkened room. You can see how much dust is in the air. You can clean up the dust with a comb that has been electrically charged by rubbing it against a wool scarf or sweater. The negatively charged comb attracts the uncharged dust particles and holds on to them.

The scrubbers that remove ash from the smoke emitted by environmentally friendly factory chimneys operate in a similar manner. Together with exhaust gases from the boiler, the ash travels through metal gratings, which are under strong direct current tension. Since charged bodies are attracted to uncharged ones, some of the ash particles are caught by the grating. Other ash charge particles get caught a little farther on at uncharged metal plates.

GIANT BOOK OF SCIENCE EXPERIMENTS

40. High voltage

Place a flat baking tray on a dry glass. Rub a balloon vigorously on a wool sweater and then place it on the tray. If you put your finger near the edge of the tray, a spark jumps across.

The spark is produced as the charge between the metal and your finger equalizes. Although the spark is discharged with several thousand volts, it is just as harmless as the sparks produced when you comb your hair. An American scientist calculated that a cat's fur must be stroked 9,200,000,000 times to produce a current sufficient to light a 75-watt bulb for a minute.

41. Lightning rod in the hand

When you walk on a synthetic-fiber carpet, shoe soles made of rubber absorb electrons removed from the fibers. As the charge builds up, some of the electrons in your body are pushed towards your fingertips and when you touch a grounded object, they will jump away from you as a small electrical spark. Even though the tension of such sparks contains several thousand volts, they are harmless due to their small amount of current. If you don't like that shock that the tiny sparks deliver, you can shunt them via a key or something similar. Like a lightning rod on the roof, which diverts the lightning from a thunder cloud away from a house to the earth, the energy passing between the body and the grounded object unloads itself through the well conducting metal in your hand.

                                       GIANT BOOK OF SCIENCE EXPERIMENTS

42. Flash of lightning

Place a metal cake slicer or similar object on a dry glass, and on it a piece of hard Styrofoam plastic which you have rubbed well on your sweater. If you hold your finger near the handle of the cake slicer, a spark jumps across.

When the negatively charged plastic is placed on the slice, the electrons in the metal are repelled to the end of the handle. When the spark occurs, the charge between it and your finger becomes equalized. Plastic materials can become strongly charged. In warehouses, for example, metal stands for rolls of plastic are earthed because otherwise they often spark when they are touched by the personnel.

43. Electric light

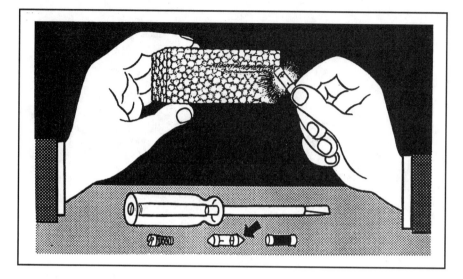

See if you can find a voltage tester, shaped like a screwdriver. In its handle there is, among other things, a small neon tube which you can easily remove. Hold one metal end firmly and rub the other on a piece of hard Styrofoam plastic, which is often used for packing or insulation. The lamp begins to glow as it is rubbed to and fro. You can see this particularly clearly in the dark.

Since the plastic is soft, its layers are rubbed against one another by the movement of the lamp and become strongly charged with electricity. The electrons collect on the surface, flow through the core of the tiny lamp, which begins to glow, and into your body.

The ancient Greeks had already discovered that amber attracted other substances when it was rubbed. They called the petrified resin "electron."

54

44. Effects of cathode rays

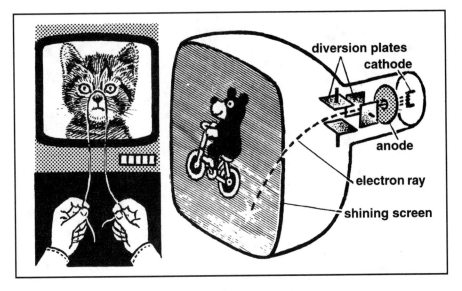

diversion plates
cathode
anode
electron ray
shining screen

When the screen of a television is turned on, it attracts more than dust. If you hold a cotton thread about a yard long in front of it, the thread levitates horizontally. Why?

The cathode ray (a stream of electrons sent out from the cathode, a glowing wire at the back of the TV tube) causes the attraction. First, the beam of electrons is attracted towards the anode, a metal disk with a circular hole, which focuses the beam in such a way that it hits the screen and glows as a small dot. Guided by charged plates, the ray can be shifted rapidly in the vertical and horizontal planes. As the beam moves, it lights the 800 dots in each of 525 lines with a different brightness, to create a new image 30 times a second.

The electrons, which hit the screen, charge it electrostatically like a pane of glass rubbed with a silk cloth. The negatively charged screen then attracts uncharged things.

45. Shooting puffed rice

Charge a plastic spoon with a woollen cloth and hold it over a dish containing puffed rice. The grains jump up and remain hanging on the spoon until suddenly they shoot wildly in all directions.

The puffed rice grains are attracted to the negatively charged spoon and cling to it for a time. Some of the electrons pass from the spoon into the puffed rice, until the grains and the spoon have the same charge. And because like charges repel one another, the puffed rice grains fly away from the spoon.

46. Field lines

Lay a sheet of drawing paper over a magnet—of course you already know how to make a magnet—and scatter iron filings on it. Tap the paper lightly, and a pattern forms.

The filings form curved lines that reveal the direction of the magnetic force. You can make the pattern permanent. Carefully dip paper into melted candle wax and let it cool. Scatter the iron filings on it. If you hold a hot iron over the paper after the formation of the magnetic lines, the field lines, the pattern will be fixed.

47. The earth's magnetic field

Hold a soft iron bar pointing to the north and sloping downwards, and hammer it several times. It will become slightly magnetic.

The earth is surrounded by magnetic field lines, which meet the earth in North America and Great Britain at an angle between 60° and 80°. When the iron is hammered, some of its magnet particles are aligned by the earth's magnetic field lines so that they point to the north. In a similar way, tools sometimes become magnetic for no apparent reason. If you hold a magnetized bar in an east–west direction and hammer it, it loses its magnetism.

48. Magnetic or not?

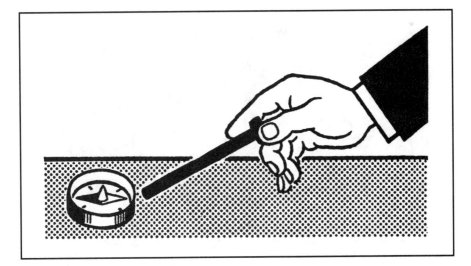

Many iron and steel objects are magnetized without one's realizing it. You can detect this magnetism with a compass.

If a rod is magnetized, it must, like the compass needle, have a north and south pole. Since two unlike poles attract and two like poles repel, one pole of the needle will be attracted to the end of the rod and the other repelled. If the rod is not magnetized, both poles of the needle are weakly attracted to the end.

49. Compass needle

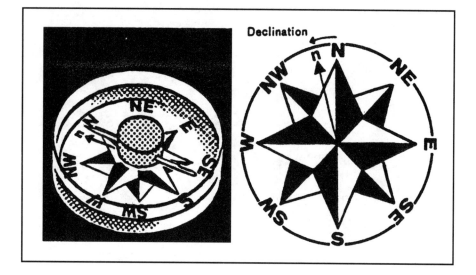

Stroke a sewing needle with a magnet until it is magnetized and push it through a cork disk. Put the needle into a transparent plastic lid containing water and it turns in a north–south direction. Stick a paper compass card under the lid.

The needle points towards the magnetic North Pole of the earth. This lies in northern Canada, and is not to be confused with the geographical North Pole, around which the earth rotates. The deviation (declination) of the magnetic needle from true north is 8° in London and 15° in New York (in a westerly direction) and 1° in Chicago and 15° in Los Angeles (in an easterly direction).

50. Dip to the pole

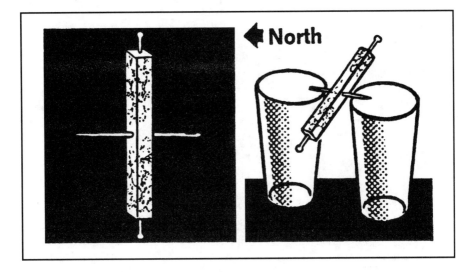

North

Magnetize two steel pins so that their points attract each
other strongly. Push them into the ends of a piece of foam
plastic about as thick as a pencil and push a sewing needle
through the middle. Balance the assembly between two
glasses (by shifting the pins and pulling off pieces of
plastic). If you allow this compass to swing in a north–south
direction, it will come to rest with the end facing north
sloping downwards.

The compass needle comes to rest parallel to the
magnetic field lines that span the earth from pole to pole.
This deviation (dip) from the horizontal is 67° in London,
72° in New York, 60° in Los Angeles, and 90° at the
magnetic poles of the earth.

51. Magnetic ducks

Make two ducks from paper folded over and glued and push a magnetized pin into each one. Set the pins so that the two poles that will be attracted are placed in the ducks' beaks. Place the ducks on cork disks in a dish of water. After moving around, they line up with their beaks or tail tips together in a north–south direction.

The ducks approach each other along the magnetic field lines. Their movement is caused by different forces: the attraction of unlike magnetic poles, the repelling effect of like poles, and the earth's magnetism.

52. Atmospheric pressure

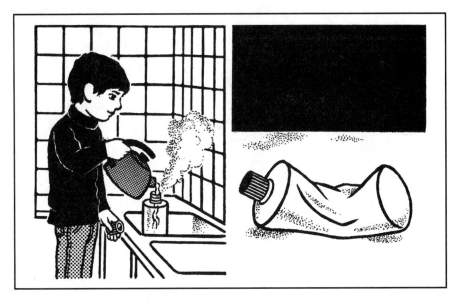

Try rinsing a plastic bottle with very hot water, closing it immediately so that it is airtight, and then putting it into the refrigerator. It becomes flat. How does that happen?

The heat from the water causes the air in the bottle to expand by about one-third, so that part of it escapes. In the process of cooling off, the air contracts again, creating a low pressure in the bottle. Because the pressure of the outside air is greater, it compresses the bottle until the air pressure inside is equal to the pressure of the outside air.

The enormous weight of the atmosphere, which surrounds the earth, becomes evident. It presses with 1 atmosphere, an amount equivalent to a little more than 14 pounds on each square inch of the bottle's surface.

53. Diving bell

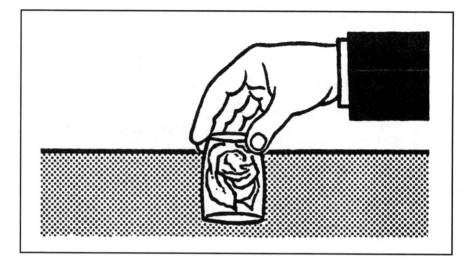

You can immerse a pocket handkerchief in water without it getting wet: stuff the handkerchief firmly into a glass and immerse it upside down in the water. Although air is invisible, it nevertheless consists of molecules of oxygen, nitrogen, and other gases which fill the available space. So air is also enclosed in the upturned glass, and it stops the water from entering. If, however, you push the glass deeper, you will see that some water does enter, due to the increasing water pressure, which compresses the air slightly. Diving bells and caissons, used underwater, work on the same principle.

54. Balloon in the bottle

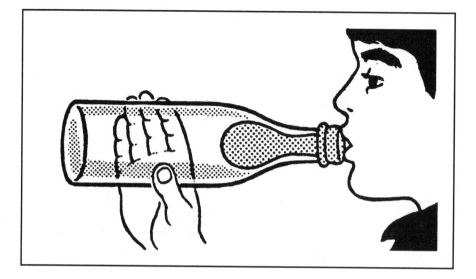

Do you believe that it is always possible to blow an ordinary balloon right up? You will be surprised. Push a balloon into a bottle and stretch its mouthpiece over the opening. Blow hard into the balloon. It is only possible to stretch the rubber before your breath runs out.

As the pressure of the air in the balloon increases, so does the counterpressure of the air enclosed in the bottle. It is soon so great that the breathing muscles in your chest are not strong enough to overcome it.

55. Air lock

Place a funnel with a narrow spout in the mouth of a bottle and seal it with modeling clay so that it is airtight. If you pour some water into the funnel, it will not flow into the bottle.

The air enclosed in the bottle prevents the water from entering. On the other hand, the water molecules at the mouth of the funnel, held together by surface tension, form a sort of skin which does not allow any air to escape. Close one end of a straw with a finger, push the other end through the funnel and lift your finger. The water can now flow into the bottle, as air escapes through the straw.

56. Hanging water

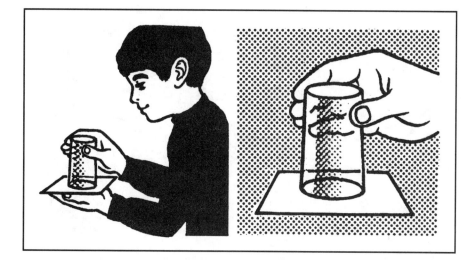

Fill a glass to overflowing with water and lay a postcard on it. Support the card with one hand, turn the glass upside down, and remove your hand from the card. It remains on the glass, and allows no water to escape.

With a glass of normal height, a weight of water of about 2 ounces presses downward on each square inch of card. At the same time, however, the pressure of air pushing upward on the postcard from below is about 100 times as great on each square inch. The air pressure holds the card so firmly against the glass that no air can enter at the side and so no water can flow out.

57. Weight of air on paper

Lay a cigar-box lid, or a similar, thin piece of wood, over the edge of a smooth table. Spread out an undamaged sheet of newspaper and smooth it firmly on the lid. Hit the projecting part of the lid hard with your fist. It breaks, without the paper flying up.

The lid is only slightly tilted when it is hit. In the space formed between the lid, newspaper, and table, the air cannot flow in quickly enough, so that there is a partial vacuum, and the normal air pressure above holds the lid still, as if it were clamped down.

58. Fountain

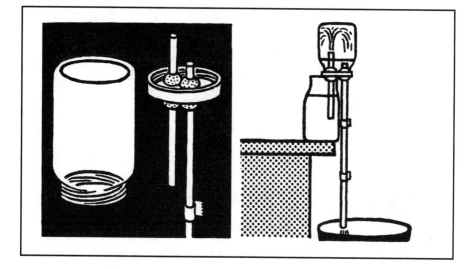

Punch two large holes in the lid of a jam jar and push a plastic straw two inches through one hole. Join three more straws together with adhesive tape and push them through the other hole. Seal the hole with modeling clay. Screw the lid to the hole jar, which should contain some water, turn it upside down, and dip the short straw into a bottle full of water. A fountain of water rises into the upper jar until the bottle is empty.

The water pours out through the long tube, lowering the air pressure in the jar. Because the air pressure outside is greater, air tries to get in and pushes the water from the bottle and up the straw.

59. Bottle barometer

Stretch a piece of balloon rubber over the mouth of a milk or juice bottle, glue a straw to the middle of it, and put a matchstick between the straw and the rim of the bottle. As the air pressure varies daily according to the state of the weather, the opposite end of the straw moves up and down.

When the air pressure is higher in good weather, the rubber is pressed inwards, and the end of the pointer rises. When the air pressure falls (as it does when overcast skies or rain are imminent), the pressure on the rubber is reduced, and the pointer falls. Because the air in the bottle will expand if it is heated, the barometer should be placed in a spot where the temperature will remain constant.

60. Shooting backwards

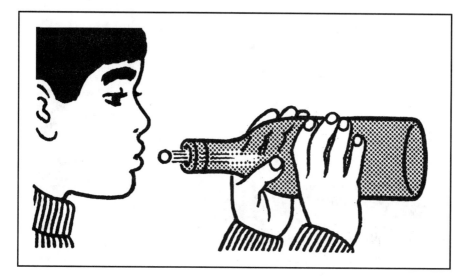

Hold an empty bottle horizontally and place a small paper ball just inside its neck. Try to blow the ball into the bottle. You cannot! Instead of going into the bottle, the ball flies towards your face.

When you blow, the air pressure in the bottle is increased, and at the same time there is a partial vacuum just inside the neck. As the pressures become equalized, the ball is driven out, as from an airgun.

61. Blowing trick

Place a playing card on a small goblet so that at the side only a small gap remains. Lay a large coin, such as a quarter, on the card. The task is to get the coin into the glass without using your hands. Anybody who does not know the trick will try to blow the coin into the gap from the side, without success. The experiment only works if you blow once quickly into the mouth of the glass. The air is trapped inside and compressed. The increased pressure lifts the card and the coin slides over it and into the glass.

62. Compressed-air rocket

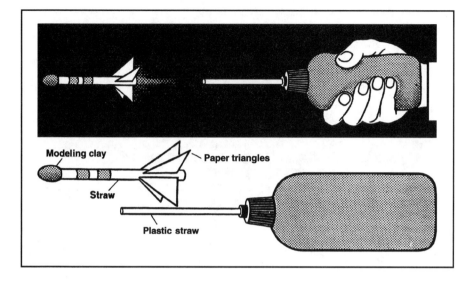

Modeling clay

Paper triangles

Straw

Plastic straw

Bore a hole through the cap of a plastic bottle, push a plastic drinking straw through it, and seal the joints between them with glue. This is the launching pad. Make the rocket from a 4-inch-long straw, which must be slightly wider so that it will slide smoothly over the first plastic straw. Glue colored paper triangles for the tail unit at one end of the straw, and at the other end modeling clay as the head. Now push the plastic tube into the rocket until its tip sticks lightly into the clay. If you press hard on the bottle the projectile will fly a distance of 10 yards or more.

When you press the plastic bottle, the air inside is compressed. When the pressure is great enough, the plastic straw separates from the plug of clay, and the air expands, propelling the projectile.

63. Compressed air in a tunnel

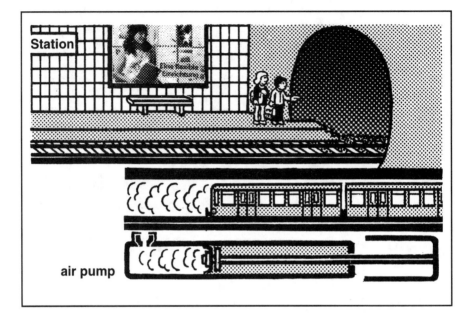

Anyone who waits for a train in a subway station near the exit of the tunnel can tell that the train is coming long before the lights and noise are detected. But how is this possible?

During the trip through the tunnel, a swell of air accumulates in front of the subway train. As soon as it departs from the station, at the next station you can feel a slight draft on your moistened finger. (The draft accelerates the evaporation of water on the finger, which cools it.) As the train approaches, a strong air flow is slowly created. The flat front of the train, which presses the air through the narrow tunnel tube, is similar to the piston in a bicycle pump.

64. Egg blowing

Place two porcelain egg cups one in front of the other, with an egg in the front one. Blow hard from above on the edge of the filled cup. Suddenly the egg rises, turns upside down, and falls into the empty cup. Because the egg shell is rough, it does not lie flat against the smooth wall of the egg cup. Air is blown through these gaps into the space under the egg, where it becomes compressed. When the pressure of the air cushion is great enough, it lifts the egg upwards.

65. Curious air currents

If you stand behind a tree trunk or a round pillar on a windy day, you will notice that it offers no protection, and a lighted match will be extinguished. A small experiment at home will confirm this: blow hard against a bottle which has a burning candle standing behind it, and the flame goes out at once.

The air current divides on hitting the bottle, clings to the sides, and joins up again behind the bottle, forming a strong eddy which hits the flame with almost as much force as blowing directly at the candle. You can put out a lighted candle placed behind two bottles in this way, if you are able to blow hard enough.

66. Bernoulli was right

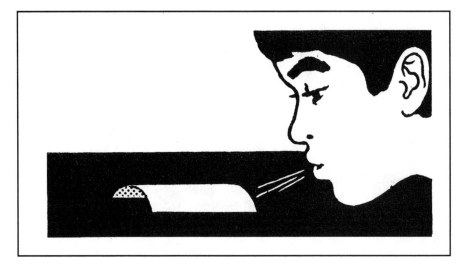

Lay a postcard bent lengthways on a table. You would certainly think that it would be easy to overturn the card if you blew hard underneath it. Try it! However hard you blow, the card will not rise from the table. On the contrary, it clings more firmly.

Daniel Bernoulli, a Swiss scientist of the eighteenth century, discovered that the pressure of a gas is lower when it is moving fast at higher speed. The air stream produces a lower pressure under the card, so that the normal air pressure above presses the card onto the table.

67. Windproof coin

Push three pins into a piece of wood as shown and lay a quarter on top of them. You can make a bet! Nobody who does not know the experiment will be able to blow the coin off the tripod.

The metal cannot be moved by air blown against its narrow edges. The air passes under the coin and reduces the air pressure below it, forcing the coin more firmly against the pins. But if you lower your chin to the wood, just in front of the coin, and blow with your lower lip pushed forward, the air hits the underside of the coin directly and lifts it off.

68. Trapped ball

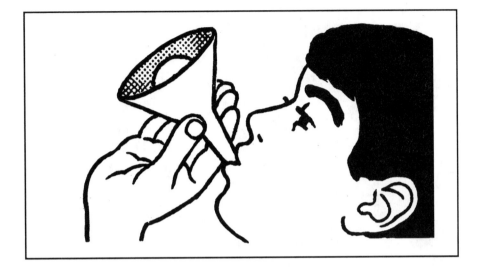

Place a table tennis ball in a funnel, hold it with the mouth sloping upwards, and blow as hard as you can through the spout. You might not believe it, but nobody can blow the ball out.

The air current does not hit the ball with its full force, as one would assume. It separates and pushes around the ball where it rests on the funnel. At these points the air pressure is lowered according to Bernoulli's Law, and the external air pressure pushes the ball firmly into the mouth of the funnel.

69. Floating card

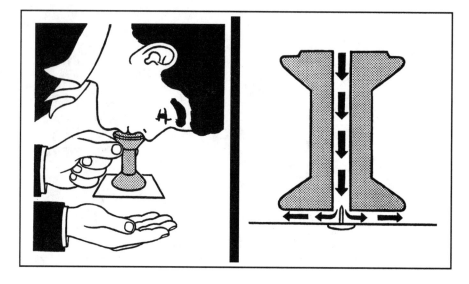

Many physical experiments seem like magic, but there are logical explanations and laws for all the strange occurrences. Stick a thumbtack through the middle of half a postcard. Hold it under a thread spool so that the pin projects into the hole, and blow hard down the hole. If you manage to loosen the card, you really expect it to fall. In fact, it remains hovering under the spool.

This surprising result is explained by Bernoulli's Law. The air current passes between the card and the spool at high speed, producing a lower pressure, and the normal air pressure pushes the card from below against the spool. Airplanes fly in a similar manner. The air flows over the arched upper surface of the wings faster than over the flat under surface, and therefore the air pressure above the wings is reduced, lifting the wings upward.

70. Flying coin

Lay a dime four inches from the edge of a table and place a shallow dish eight inches beyond it. How can you blow the coin into the dish?

You will never do it if you blow at the coin from the front—on the false assumption that the air will be blown under the coin because of the unevenness of the table and lift it up. It will only be transferred to the dish if you blow once sharply about two inches horizontally above it. The air pressure above the coin is reduced, and the surrounding air, which is at normal pressure, flows in from all directions and lifts the coin. It goes into the air current and spins into the dish.

71. Ride on air

soap bubble skin

forward drive

hovercraft   air cushion skirt

After being washed in hot water, a soapy glass that is laid on a smooth, slightly slanted, wet surface moves and glides downwards as if were floating. How can that be explained?

The air trapped in the heated glass warms up and tries to expand. But since it cannot escape under the edge, due to a fine soap bubble skin (from the detergent), the air pressure in the glass increases. It lifts up the glass and lets it glide down practically without any friction.

Hovercrafts across the English Channel ride on a similar air cushion. Large propellers, facing downward, increase the air pressure in a flexible skirt beneath it, which lifts the hovercraft over land and water. The forward drive is generated by propellers that are mounted on the roof.

72. Wind funnel

Light a candle and blow it hard through a funnel held with its mouth a little way from the flame. You cannot blow out the flame; on the contrary, it moves towards the funnel.

When you blow through the funnel the air pressure inside is reduced, and so the air outside enters the space through the mouth. The air sweeps along the funnel walls. If you hold the funnel with the edge directly in front of the flame, it goes out. Of course, if you blow at the candle with the funnel reversed, the air is compressed in the narrow spout, and extinguishes the flame immediately on exit.

73. Explosion in a bottle

Throw a burning piece of paper into an empty milk or juice bottle and stretch a piece of balloon rubber firmly over the mouth. After a few moments, the rubber is sucked into the neck of the bottle and the flame goes out.

During combustion the air is heated and expands, pushing some of it out of the bottle. After the flame goes out, the remaining air in the bottle cools and contracts, compressed by the external air pressure. The rubber is stretched so much that the pressure equalizes completely only if you prick the bubble with a pin, causing a loud pop.

74. Twin glasses

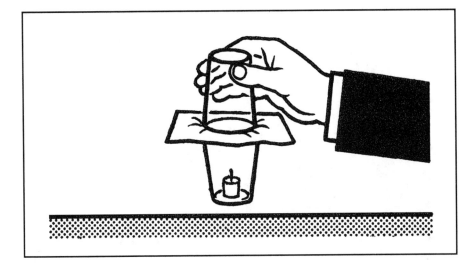

Light a candle stump in an empty glass, lay a sheet of damp blotting paper over the top, and invert a second glass of the same size over it. After several seconds the flame goes out and the glasses stick together.

During combustion the oxygen in both glasses is used up—the blotting paper is permeable to air. Therefore, the pressure inside is reduced and the air pressure outside pushes the glasses together.

75. Coin in the well

Place a coin in a dish of water. How can you get it out, without putting your hand in the water or pouring the water from the dish? Put a burning piece of paper in a glass and invert it on the dish next to the coin. The water rises into the glass and uncovers the coin.

During combustion the carbon contained in the paper, together with other substances, combines with the oxygen in the air to form carbon dioxide. The gas pressure in the glass is reduced by the expansion of the gases on heating and contraction on cooling. The air pressure from outside pushes the water into the glass.

76. Bottle ghost

An empty bottle that has been stored in a cool place has a ghost in it! Moisten the rim of the mouth with water and cover it with a coin. Place your hands on the bottle. Suddenly the coin will move as if by a ghostly hand.

The cold air in the bottle is warmed by your hands and expands, but is prevented from escaping by the water between the bottle rim and the coin. However, when the pressure is great enough, the coin behaves like a valve, lifting up and allowing the warm air to escape.

77. Expanded air

Pull a balloon over the mouth of a bottle and place it in a saucepan of cold water. If you heat the water on a stove, the balloon inflates.

The heat causes the air molecules in the bottle to whirl around faster in all directions, so that they move farther apart and the air expands. This causes an increase in pressure, which causes the balloon to blow up. If you take the bottle out of the saucepan, the air gradually cools down again and the balloon collapses.

78. Thermometer

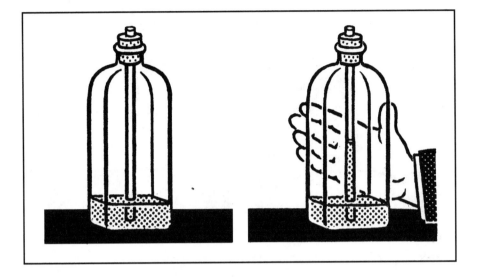

Pour some colored water into a bottle. Push a drinking straw through a hole bored in the bottle's cork so that it dips into the water. Seal the cork with glue. If you place your hands firmly on the bottle for a period of time to warm it (or place it in hot water), the colored water rises up the straw.

The air enclosed in the bottle expands on heating and presses down on the water surface. The displaced water escapes into the straw and shows the degree of heating by its position. You can glue a scale on the side of the bottle.

79. Hot-air balloon

Roll a paper napkin (not too soft material) into a tube and twist up the top. Stand it upright and light the tip. While the lower part is still burning, the ash formed rises into the air. But be careful where it lands! The air enclosed by the paper is heated by the flame and expands. The light, balloon-like ash residue experiences a surprising buoyancy because the hot air in the balloon is lighter than the surrounding cool air. Very fine napkins are not suitable for the experiment because the ash formed is not firmed enough.

80. Expanding metal

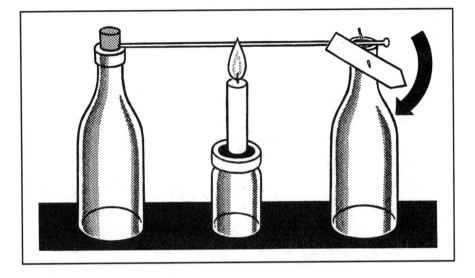

Take an empty, corked bottle, push as long an aluminum knitting needle as you can find into the bottle's cork and let the other end rest on the mouth of a second, uncorked bottle. Glue a paper arrow onto a sewing needle, making sure that it is balanced, and place it between the knitting needle and the neck of the bottle as shown. Place a candle so that the tip of the flame touches the middle of the needle and watch the arrow.

The arrow turns quite quickly to the right because the knitting needle expands on heating, like other substances. With an ordinary steel knitting needle the arrow would only turn a little, because steel only expands half as much as aluminum. Also, the longer the knitting needle, the more the arrow will move. If you take the candle away from the knitting needle, the arrow moves back.

81. Color and heat

Place two candlesticks in front of a window, one made out of white glass, the other out of black glass. At the end of a sunny day, the candle in the white candleholder has not changed, but the candle in the black candleholder has bent downwards, strangely enough, in the direction of the sun. How much an object warms up in the sun depends on its reflectivity and color. The white candleholder reflects almost all the light striking it and warms up only a little; the black one absorbs most of the light and warms up more. Glass is a poor conductor of heat, so the side of the candleholder which faces the sun is the first to reach the temperature at which wax becomes soft. On this side, therefore, the candle loses its firmness and bends down.

82. Exploding stone

You can explode large stones in the winter quite easily. Look for a flint that is well frozen through and pour boiling water over it. It breaks apart with cracks and bangs. The explosive effect is caused by the outer layers heating and expanding faster than the center. The resulting tension causes the stone to burst. In the same way, thick-walled glasses may explode if you pour hot liquids into them. Glass conducts heat poorly, so that the inner and outer layers of glass expand by different amounts.

83. Tension in glass

If you pour boiling water in a glass, it can easily crack. But this can be avoided, by placing a silver spoon in the glass. Why? Glass is a poor conductor of heat. In the case of a glass with thick walls, it takes a while for the heat to move from the inside to the outside. Due to the uneven warming, the glass layers expand to different degrees, and tension is created that can crack the glass. Silver conducts heat 350 times better than glass. The spoon therefore absorbs much of the heat at once and transfers it up the handle. The tension in the glass is reduced, because now the warming and expansion of the glass layers can take place more slowly and evenly.

84. A clear case

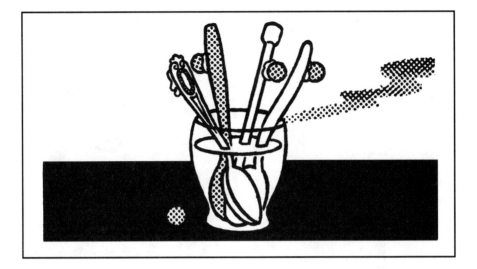

Put spoons made of steel, silver, and plastic and a glass rod into a glass. Fix a dry pea at the same height on each handle with a dab of butter. In which order will the peas fall if you pour boiling water into the glass?

The butter in the silver spoon melts very quickly and releases its pea first. The peas from the steel spoon and the glass rod fall next, while the pea on the plastic spoon does not move. Silver is by far the best conductor of heat, while plastic is a very poor conductor, which is why saucepans often have plastic handles.

85. Scented-coin trick

Three different coins lie in a plastic dish. You close your eyes while another person takes out one coin, holds it for several seconds in his closed hand, and puts it back. Now hold each coin briefly against your upper lip and pretend to sniff it. To everyone's astonishment, you can immediately detect which coin was taken from the dish. Since metals are very good conductors of heat, the coin warms up quickly in the hand. But plastic is a poor conductor, so hardly any heat is lost to the dish when the coin is put back. The upper lip is particularly sensitive and reveals the smallest temperature difference in the coins, so that you can detect the right one immediately. Before the trick is repeated it is a good idea to lay the coins on a cold stone floor or another cool surface to conduct away the heat.

86. Fire guard

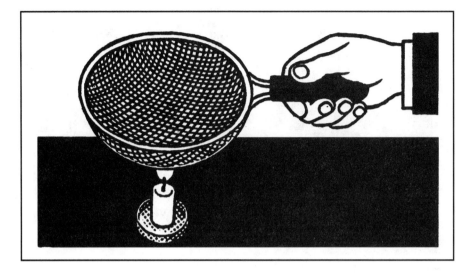

Hold a metal kitchen sieve in a candle flame. To your surprise the flame only reaches the wire net, but does not go through it.

The metal in the sieve conducts so much heat away that the candle wax vapor cannot ignite above the wire net. The flame only passes through the metal lattice if it is made to glow by strong heating. The miner's safety lamp works in the same way. A metal lattice surrounding the naked flame takes up so much heat that the gases in the mine cannot ignite.

87. Fire underwater

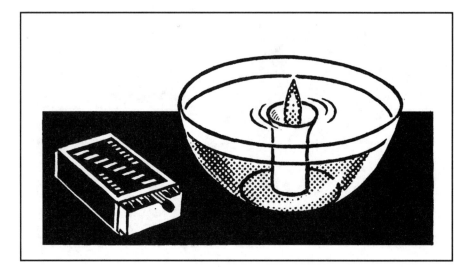

Warm the base of a candle stump and stick it in a bowl. Fill the bowl with cold water up to the rim of the candle. If you light the wick it burns until it is under the surface of the water. Then the candle flame hollows out a deep funnel. An extremely thin wall of wax remains standing around the flame, preventings the water from extinguishing it.

The water takes so much heat from the candle that its outer layer does not reach its melting point, and the wax there cannot evaporate and burn.

88. Paper saucepan

Do you think that you can boil water in a paper cup over a naked flame or in the embers of a fire? Push a knitting needle through the rim of a paper cup containing a little water, hang it between two upright bottles, and light a candle under the cup. After a little while the water boils— but the cup is not even scorched.

The water removes the heat transferred to the paper and begins to boil at a temperature of 212°F. The water does not get any hotter than that, so the paper cannot reach the temperature which is necessary for it to burn until all the water has been converted into steam.

89. Forces in steel

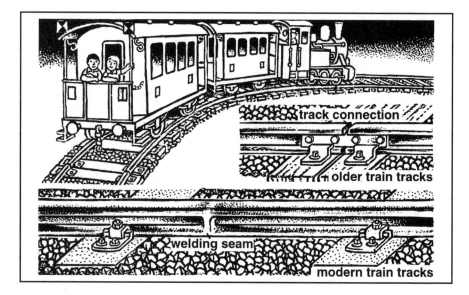

On a train ride over older tracks, you will hear the monotonous rattling of the wheels going over the track connections. In the winter, the noise is louder than in the summer. Why?

Steel expands, like any other metal, in warmth, due to increased molecular movement, and contracts again in the cold. The gaps between the tracks are therefore wider in the winter than in the summer. At a temperature difference of 90°F, the rail sections, which are 30 feet long, change in length by about a quarter of an inch.

Tracks on modern routes, which are welded together without any gaps, are thicker and welded at an average temperature. Because of their enormous weight, stronger sleeper ties, and wider gravel beds, the rails press to the side instead of expanding lengthwise.

GIANT BOOK OF SCIENCE EXPERIMENTS

90. Different body temperatures

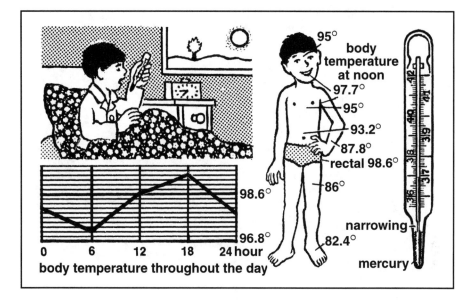

95°
body temperature at noon
97.7°
95°
93.2°
87.8°
rectal 98.6°
86°
narrowing
82.4°
mercury

98.6°

96.8°

0   6   12   18   24 hour
body temperature throughout the day

A boy wants to pretend that he has a fever. Will he succeed by breathing on the thermometer to drive up the mercury column?

The normal temperature in the inside of the body of a person (measured rectally) ranges during the day between 97.2 and 99.3°F. The breath cannot be warmer. You only perceive it as being warmer, because the temperature on the body surface is lower.

The mercury column in the thermometer, which rises as the liquid metal expands when warmed up, holds at that level because of the narrow hole in the lower part of the tube. Each time before the temperature is read again, the mercury therefore has to be struck down.

HEAT

91. Jet boat

Punch a small hole from the inside through the screw top of an aluminum tube about four inches long, and pour a little water into the tube. Fix the tube in an empty sardine can into which you have fixed three candle stumps and place the can in water. If you light the candles the water soon boils, and the jet of steam escaping from the back drives the boat.

Steam is formed in the metal tube when the water boils. Because water expands sharply as it turns to steam, it escapes at high pressure through the nozzle and causes a recoil. Do the experiment in calm weather!

92. Hovercraft

Place a pan on a hot plate and heat it well (be careful!). If you then let a few drops of water fall on the lid, you will observe a strange phenomenon. The drops are suspended in the air like hovercraft and whizz around the pan until they disappear.

On contact with the heated metal, the undersides of the water drops are instantly converted to steam. Since the steam escapes with great pressure, it lifts the drops into the air so that they move about the pan with little friction. So much heat is removed from the drops by the formation of steam that they do not even boil.

93. Weather station

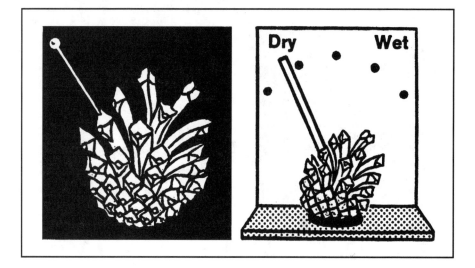

Glue a dry pinecone to a small piece of wood with sealing wax or glue. Stick a pin into one of the central scales and place a straw over it. Put the cone outdoors, protected from the rain. The straw moves according to the state of the weather.

This simple hygrometer (an indicator of humidity) is one of nature's designs. Draw a scale, recording the movements. The pinecone closes when it is going to rain, to protect the seeds from the damp. The outside of the scales absorbs the moisture in the air, swells up, and bends—a process which you can also observe with a piece of paper that is wet on one side.

94. Evaporation and condensation

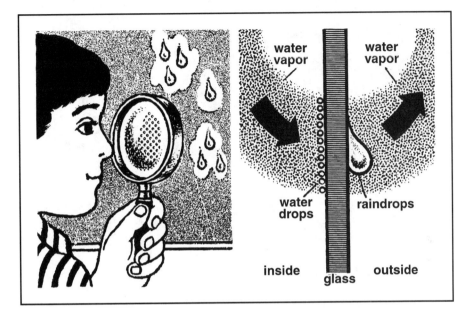

In windy weather raindrops are hanging on the outside of the window panes while the glass around the drops becomes clouded on the inside. Where does the humidity come from?

The drops on the outside of the pane evaporate in the wind. In the process of evaporating, they take away heat from the glass in the vicinity of the drops. On the inside, the warm room air, which contains a lot of water vapor, cools off, when it passes the cold spots on the glass. But since cool air cannot hold as much humidity as warm air, part of the water vapor condenses. That means it liquefies and precipitates as tiny water drops on the pane, which you can see through a magnifying glass.

EVAPORATION AND VAPORIZATION

95. Condensation cloud in a bottle

When you open a soda bottle, a small condensation cloud often forms at its neck. Why is that?

The air in the closed bottle is saturated with water vapor and is compressed by carbon dioxide gas, which is dissolved in the soda. When the bottle is opened, the pressure of the air is suddenly reduced, causing the air to cool. The air, once cooled off, cannot hold as much water vapor and discharges part of it as fine droplets. Condensation trails, or contrails, in the sky are created in a similar way. A stream of combustion gases shoots out of the engine of an airplane. They contain, among other things, water vapor, which condenses to form little droplets when the gas loses pressure and cools off.

GIANT BOOK OF SCIENCE EXPERIMENTS

96. Water from the desert

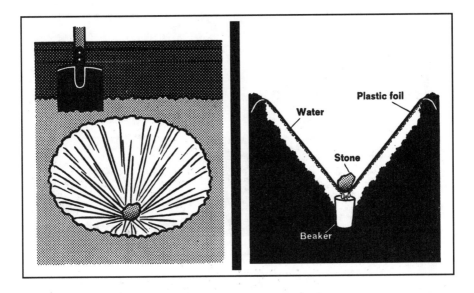

We still read in the newspapers of people dying of thirst in the desert, but many of them could help themselves in this emergency. An experiment on a small scale in a sand-box will show you how to do it.

Dig a fairly deep hole and place a beaker in the middle. Spread a large piece of transparent plastic foil, or garbage bag, flat across the hole and fix the edges firmly in the sand. Lay a small stone in its center so that it dips the foil down to the beaker in the shape of a funnel. Soon, small drops of water form on the underside of the foil. They become larger and finally flow into the beaker. The sun heats the ground under the foil, so that any moisture held in the sand evaporates until the enclosed air is so saturated that small drops of water are deposited on the cooler foil.

97. Bath game with a coin

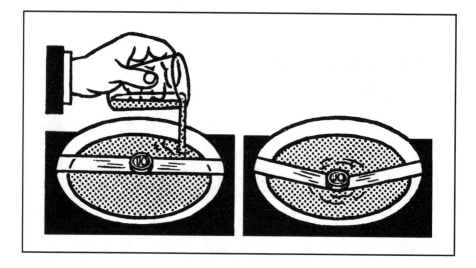

Stretch a strip of cellophane (not plastic foil), 1 inch wide, tightly over a soup plate and fasten the ends with adhesive tape. Lay on the middle of the strip an average-sized coin and pour water into the dish up to about $1/2$ inch under the coin. The coin sinks slowly and reaches the water after several minutes.

The water evaporates, the cellophane absorbs the water molecules from the air, and expands until it reaches the water. But, strangely enough, it soon begins to tighten again, and the coin rises again slowly to its original position.

98. Steamboat

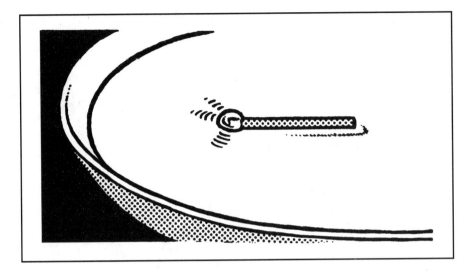

Break off the head of a match and add a drop of airplane model glue onto the end. If you place the match in a dish of water, it moves jerkily forward.

The glue contains a solvent which evaporates to give a vapor. It puffs out from the drop in invisible little clouds, giving the match a small push each time. Eventually so much of the solvent has escaped that the glue becomes solid. In a dried drop of glue you can still see the residual solvent vapor as small bubbles.

99. Where is the wind coming from?

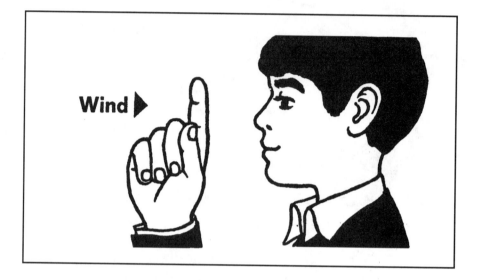

Wind ▶

Moisten your finger and hold it straight up in the air. You will notice at once that one side of the finger is cold. This is the direction from which the wind is coming.

Heat is used up when a liquid vaporizes or evaporates. The wind accelerates the evaporation of the moisture on the finger and you will notice, even with a weak air current, the greater heat loss on the side facing the wind. Anybody who keeps on a wet bathing suit after a swim will shiver even on a warm day. The water takes heat from the body as it evaporates.

100. Producing cold

With a rubber band, attach a wad of cotton over the mercury bulb of a room thermometer. Note the temperature, dampen the cotton with rubbing alcohol, and whirl the thermometer around on a string for a time. The temperature drops considerably. The alcohol evaporates quickly (much faster than water) and so uses up heat. The air flow caused by whirling the thermometer accelerates the process and the heat consumption rises. In a refrigerator a special liquid evaporates within a closed system. The large amount of heat needed for this is taken from the food compartment.

COLD AND ICE

101. Column of ice

Place an open ink bottle filled to the brim with water in the freezer. Soon a column of ice will stick up out of the bottle.

Water behaves oddly: when warm water cools, it contracts; but if the temperature falls below 39°F, it begins to expand again. At 32°F it begins to freeze, and in doing so increases its volume by one-eleventh. This is the reason why the ice sticks out of the bottle. If you had closed it, it would have cracked. In winter water can burst pipes, and when it collects under asphalt roads and freezes, it produces cracks and potholes.

102. Iceberg

Place a cube of ice in a glass and fill it to the brim with water. The ice cube floats and partly projects from the surface. Will the water overflow when the ice cube melts?

The water increases its volume by one-eleventh when it freezes. The ice is therefore lighter than water, floats on the water surface, and projects above it. It loses its increased volume when it melts and exactly fills the space which the ice cube took up in the water. This is why icebergs are so dangerous to ships; one only sees their tips above the water.

103. Cutting through ice

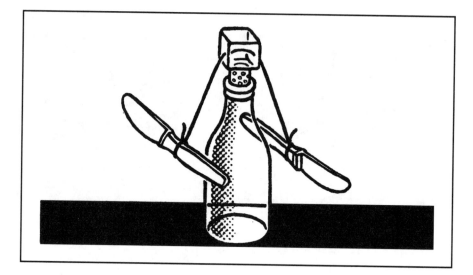

Place an ice cube on the cork of a bottle. Fix two objects of equal weight on a piece of wire, hang the wire over the ice, and place the experiment outside in frosty weather. After a certain time the wire will have cut through the ice without dividing it.

This trick of nature is explained by the fact that ice melts when it is subjected to pressure. So water is formed where the wire is pushing down, while it immediately freezes again above it. Ice skating is only made possible by slight melting of the ice under the moving skate blade, which reduces the friction.

104. Ice hook

Can you lift an ice cube from a bowl of water with a match?
A trick makes it quite easy: place the match on the ice cube
and scatter some salt over it. In a few minutes the match is
frozen solid, and you can lift if together with the ice cube
from the dish.

Salt water does not freeze as easily as ordinary water, and
scattering salt on ice makes it melt. The salt grains on the
ice cube also do this. However, when a substance melts,
heat is consumed at the same time. This heat is taken from
the moisture under the match, where no salt fell, in this
case—and it freezes.

105. Water temperatures and the wind

The water temperature at the beach can change considerably from one day to the next. It not only depends on the temperature of the day, but also on the wind direction. Is the water warmer when the wind blows from the water, or from the land out to sea.

Because water is most dense at 39°F, the more it warms up, the more it expands and the lighter it becomes. Consequently, the water layer close to the shore, which is warmed up by the sun, remains on the surface. When the wind blows from the water, the warm layer is dammed up against the shore. But if the wind shifts and blows toward the sea, it drives away the warm layer, and cool water from deep currents rises along the shore.

106. Temperature in winter ponds

snow
ice
33.8°F
35.6°F
37.4°F
39.2°F

A girl took a goldfish out of a pond at the beginning of winter and kept it outside in a tub. After the first frost, the girl found the fish almost dead, lying on its side. She wondered why the fish was doing so poorly when it survived the winter just fine in the pond.

The water, with a temperature barely above the freezing point, is too cold for the fish. Since in the tub it is exposed to the cold air from all sides, the water cannot hold at 39°F, which is the optimal winter temperature for the fish. In a deep pond or lake, the coldest water layers hover directly underneath the ice, because they are lighter than the deep water layers near the bottom, which have a temperature of 39°F. This warmer water has the highest density, so it remains at the deepest spots, where the fishes hibernate.

107. Explosive force of the ice

After every harsh winter, bumps—above which the asphalt crumbles—appear in some roads. When are these bumps worse: when there is a prolonged cold spell, or when there are alternating periods of freezing and thawing?

Through fine, hairline cracks in the asphalt, water gets under the street surface and gathers in hollow spaces. When the water freezes, it enlarges its volume by one-eleventh, and the ice presses the asphalt surface upwards. After a thaw, there is one-eleventh more room for new water created in the enlarged hollow spaces, which expand again when it freezes again. That way, the bumps get bigger each time when there is a change in the weather.

108. Icicles

On the snow-covered roof of a stable icicles form in the winter. Sometimes they are long, thin icicles, other times they are short and thick. What is the reason for the different shapes?

The radiating body warmth of the stable animals, which remains the same, causes the lowest layer of snow on the roof to melt. The meltwater is insulated against the cold by the layers of snow above, and trickles down over the edge of the roof with a temperature slightly above freezing. When the water hits the outside air it loses the little warmth that it has and freezes, sometimes slowly, sometimes fast, depending on how cold it is. The drops, which form icicles layer by layer, travel farther in mild cold before freezing (A); in extreme cold the drops travel only a short distance before freezing (B).

109. Dirt in snowflakes

Each crystal that forms a snowflake contains at its center a tiny speck of dust. The wind lifted it up from the earth and carried it high into the clouds where moisture condensed on it and immediately froze to ice. Falling to earth again, it collected more water molecules which, depending on the degree of moisture, formed crystals in the shape of needles, stars, or little leaves of ice. The grime which you see on top of melting snow indicates the amount of dust which the crystals collected on their journey to earth.

GIANT BOOK OF SCIENCE EXPERIMENTS

110. Forces in a puddle

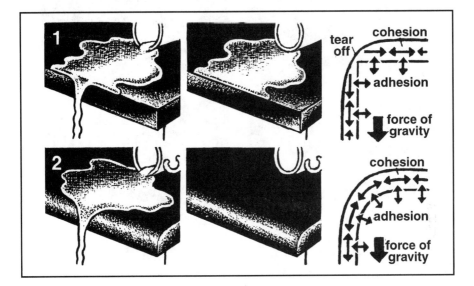

When water is spilled on a kitchen table with a smooth plastic surface, how much water runs off depends on the edge. Over a right-angled edge, the only water that runs down is that which spills over immediately after pouring; it then breaks off (1). The water which is running over a rounded edge pulls the entire puddle with it (2). Why is that?

Flowing water follows the force of gravity. On a sharp table-edge, the cohesion of the water, the attractive power between the water molecules, is interrupted. Over a rounded edge, the cohesion remains, and it is even stronger than the adhesion, the attraction between the molecules of the water and the plastic. Therefore, all of the water pours off, and the tabletop is soon dry.

111. String of pearls

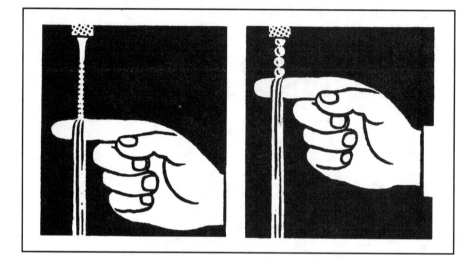

Let a fine jet of water pour on a finger held about two inches under the tap. If you look carefully, you will see a strange wavelike pattern in the water above your finger. If you bring your finger closer to the tap, the waves become continuously more ball-shaped, until the water jet resembles a string of pearls. The flow is so strongly obstructed by the finger that because of its surface tension—the force which holds the water molecules together—it backs up and forms round droplets. If you take your finger farther away from the tap, the falling speed of the water becomes greater, and the drop formation is less clear.

GIANT BOOK OF SCIENCE EXPERIMENTS

112. Water knots

With a small nail, pierce an empty 2-lb. can five times just above the lower edge. The first hole should be just over an inch from the fifth. Place the can under a running tap, and a jet will flow from each hole. If you move your finger over the holes, the jets will join together.

Water molecules are attracted to one another, producing a force called surface tension which pulls the outermost water molecules together tightly. It is this force that holds a water droplet together. In our experiment the force is made especially noticeable as it diverts the jets into sideways arcs and knots them.

113. Mountain of water

Fill a glass just full with tap water, without any overflowing. Slide coins carefully into the glass, one after the other, and notice how the water curves above the glass.

It is surprising how many coins you can put in without the water spilling over. The "water mountain" is supported by surface tension, as though it is covered by a fine skin. Finally, you can even shake the contents of a salt shaker slowly into the glass. The salt dissolves without the water pouring out.

114. Ship on a high sea

Place a quarter on a table, and on one side of the coin a small cork disk. How can you move the cork to the exact center of the coin without touching it?

Pour water on the coin—one drop at a time so that it does not spill over—to form a water mountain over the surface. At first the force of gravity holds the cork on the edge of the slightly curved water surface. If you pour on more water, the pressure of the water on the edge increases, while it remains constant on the top. So the cork moves up the hill to the middle, which is the region of lowest pressure.

115. Floating metal

Fill a bowl with tap water. Place small metal objects on blotting paper and lower it carefully into the dish with a fork. After a time, the saturated blotting paper sinks, but the small objects remain floating.

Since metal is heavier than water, it should fall to the bottom. The liquid molecules are held together so strongly by a mysterious force, the surface tension, that they prevent the objects from sinking. Surface tension is destroyed by soap.

GIANT BOOK OF SCIENCE EXPERIMENTS

116. Watertight sieve

Fill a milk or juice bottle with water and fasten a piece of wire screen about two inches square over its mouth with a rubber band. Place your hand over the top and turn the bottle upside down. If you take your hand away quickly, no water comes out. Where water comes into contact with air, it pulls together as though it were forming a skin, because of its surface tension. Each opening in the wire screen is so well sealed that air can neither flow in nor water flow out. This also occurs with the fine holes of tent material, which is made water-repellent by impregnation, and raindrops cannot get through because of their surface tension.

117. Rope trick

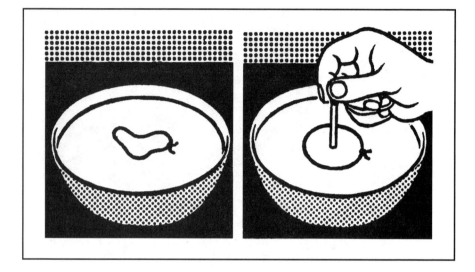

Knot a piece of string into a loop and allow it to float in a bowl of water. If you dip a match into the middle of the irregularly shaped loop, it immediately becomes circular.

The match has this magic power because it was previously dabbed with a little dish-washing liquid. This spreads in all directions when the match is dipped into the water and penetrates between the water molecules, which were held together like a skin by surface tension. This "water skin" breaks in a flash, outwards from the place where the match is dipped in. The water molecules which are made to move push against the loop and make it rigid.

118. Match boat

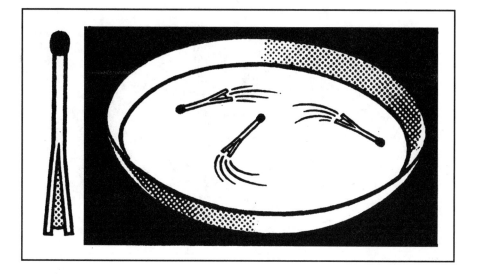

Split a match slightly at its lower end and smear some soft soap into the slit. If you place the match in a dish of tap water, it moves forward.

The soap destroys the surface tension of the water by degrees as it gradually dissolves. This causes a backward movement of the water particles, which produces as a reaction a forward movement of the match.

119. Soap trick

Make a rectangular frame about 1 × 3 inches out of thin wire. Place a straight piece of wire loosely over the center. Dip the whole thing into dish-washing liquid, so that a fine film stretches over it. If you pierce through one side, the piece of wire rolls backwards to the other end of the frame. The molecules in the liquid attract one another so strongly that the soap film is almost as elastic as a blown-up balloon. If you break the cohesion of the molecules on one side, the force of attraction on the other side predominates, and the remaining liquid is drawn over and the wire rolls with it.

GIANT BOOK OF SCIENCE EXPERIMENTS

120. Soap bubbles

In each plastic detergent bottle that is thrown away there are still a thousand soap bubbles! Cut off the lower third of an empty detergent bottle and mix 10 teaspoonfuls of water with the detergent remaining in it. Bore a hole in the cap, push a straw through it, and a match into the nozzle. Put some of the liquid into a pipe and blow!

The molecules in the soap bubbles are compressed from outside and inside by surface tension. They hold together so strongly that they enclose the air flowing from the pipe and so take on the shape with the smallest possible surface that will enclose the air—which is a sphere.

121. Visible pressure differences

Put some dishwashing liquid into a bottle of water. It gurgles in the bottle when it is poured out. As water runs out and air streams in, bubbles are created out of a fine liquid skin. Why do the smaller bubbles always arch into larger ones?

The smaller the bubble, the thicker its skin, and the stronger its surface tension, the attractive power between the liquid molecules in the surface of the air bubbles. Because of the higher tension the air trapped in small bubbles is compressed to a slightly higher pressure than in large ones. The small bubbles therefore keep their spherical shape. In the case of air balloons, it is exactly the opposite: the larger they are blown up, the stronger the air pressure is inside, due to the elastic tension of the rubber skin.

122. Bag of wind

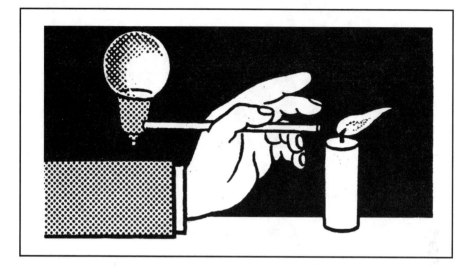

Slowly blow a large soap bubble on the end of a pipe and trap it by closing the end of the straw with your finger. Hold it near a candle flame and then take your finger away. The flame leans to the side, while the soap bubble becomes smaller and vanishes.

Although a soap bubble film is generally less than one-thirty-thousandth of a inch thick, it is so strong that the air inside is compressed. When the end of the straw is released, the liquid particles contract to form drops again because of the surface tension, and thus push the air out.

123. Water lily

Cut out a flower shape from smooth writing paper, color it with crayons, and fold the petals firmly inwards. If you place the paper flower on water you will see the petals open in slow motion.

Paper consists mainly of plant fibers, which are composed of extremely fine tubes. The water fills these so-called capillary tubes, causing the paper to swell and the petals of the flower to open up, like the leaves of a wilting plant when it is placed in water.

GIANT BOOK OF SCIENCE EXPERIMENTS

124. Game of chance

Fill a tall jar with water, place a small glass in it, and try to drop coins into the glass. It is very surprising that however carefully you aim, the coin nearly always slips away to the side.

It is very seldom possible to get the coin to drop straight into the water. The very smallest tilt is enough to cause a greater resistance of the water on the slanting underside of the coin. Because its center of gravity lies exactly in the middle, it turns easily and drifts to the side.

125. Sloping path

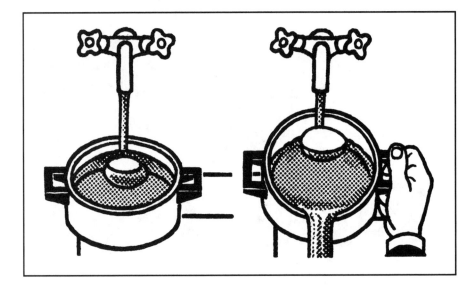

If you cool a boiled egg in the usual way in a saucepan under the water tap, you can make a surprising discovery. Hold the pan so that the water runs between the egg and the rim. If you now lean the saucepan to the other side, the egg does not, as you would expect, roll down the bottom of the saucepan but stays next to the stream of water.

By Bernoulli's Law the pressure of a liquid or a gas becomes lower with increasing speed (see experiments 66–72). In the stream of water between the rim of the pan and the egg there is reduced pressure, and the egg is pressed by the surrounding water, which is at normal pressure, against the pan.

126. Climbing liquids

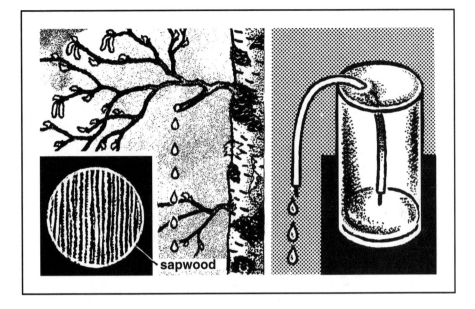

sapwood

If you cut off sprouting branches from a birch tree in the spring, a clear juice begins to trickle out of the spot where it was cut. How can the "bleeding" of the tree be explained?

In the spring, an especially large amount of water with dissolved nutrients is channeled from the roots to the buds. The transport takes place because of pressure in the roots, but also through capillary action: the molecules of the water and the wood attract each other, and because of that the juice climbs up in the tube-like pores in the sapwood.

Experiment: Stick a long, thin knitting needle through a plastic straw and bend it into a U-shape. When you hang it over the edge of a glass filled with water, the water climbs, due to the capillary action, beyond the bend, and trickles out.

127. Water pressure in little tubes

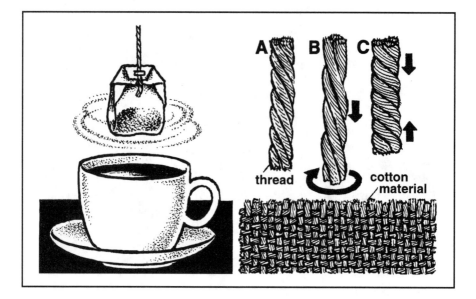

After you brew some tea, pull the teabag out of the cup. The bag turns for a while. Why?

The thread consists of spirals of twisted cotton fibers, fine tubes, which absorb water (A). Since they become thicker, a pressure is created between the spirals, and the thread twists itself apart and becomes longer (B). So when you remove the used bag the twists unwind and spin the bag. In a cotton material, the threads are so tightly interwoven that they cannot turn around when they get wet. Their spirals press outwards towards the sides so that the threads become thicker and shorter, and the material shrinks (C).

128. Loss of weight

Tie a stone by means of a thread to a spring balance and note its weight. Does it weigh the same if you hang the stone in a jar of water?

If you lift up a large stone under the water when you are bathing you will be surprised at first by its apparently low weight. But if you lift it out of the water, you will see how heavy it actually is. In fact an object immersed in a liquid (or in a gas) loses weight. This is particularly obvious with a floating object. Look at the next experiment.

129. Archimedes' principle

Fill a container up to the brim with water and weigh it. Then place a block of wood on the water. Some of the liquid will spill out. Weigh the container again, to find out if the weight has changed.

The weight remains the same. The water spilt out of the container weighs exactly the same as the whole block of wood. In about 250 B.C., the famous mathematician Archimedes discovered that a body immersed in a liquid loses as much weight as the weight of liquid displaced by it. This apparent loss of weight is called buoyancy.

130. Water puzzle

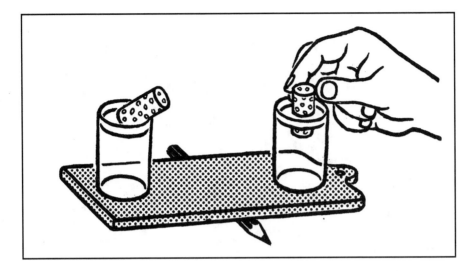

On a small tray or a wooden ruler resting on a six-sided pencil, balance two jars filled with water. What happens if you immerse a cork into the water in one jar, while placing a cork of the same size on the water in the other jar? Does one side of the balance become heavier, and if so, which side?

The balance leans to the side where you immerse the cork in the jar. That is, this side increases in weight by exactly as much as the weight of water displaced by the cork. The other jar only becomes as much heavier as the weight of the cork itself.

131. Mysterious water level

Place a heavy coin in a match box and float it in a glass of water. Mark the level of the water on the side of the glass. Will it rise or fall if you take the coin from the box and lower it into the water? Just think about it first!

The water level falls. Since the coin is almost ten times heavier than water, the box containing the coin also displaces, because of its larger volume, ten times more water than the coin alone. The coin, in spite of its greater weight, takes up only a small volume and so displaces only a small amount of water.

132. Air in a chicken egg

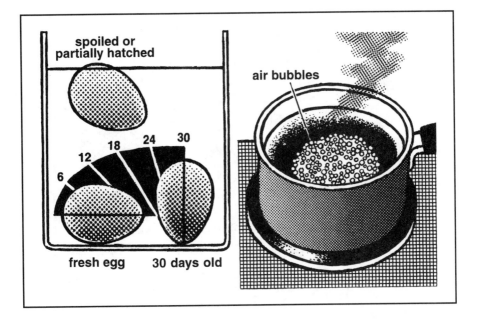

If you place a chicken egg into a pot with water, you can estimate the age of the egg by the angle of its longitudinal axle to the bottom of the pot. That is to say, a fresh egg lies horizontally in the water, a 30-day-old egg stands on its point. How can the difference be explained?

The older an egg is, the larger its air chamber becomes. The air chamber is located in the thick end, so this becomes lighter with age. This happens because a constant exchange of air takes place through the eggshell and water slowly evaporates from inside. Oxygen is needed, as the chicken embryo develops, so the eggshell has tiny pores. When the egg lies in hot water and is heated up, the expanding air inside emerges from the holes, forming small bubbles.

BUOYANCY

133. Suspending an egg

Fill a jar halfway with water and dissolve plenty of salt in it. Now add as much water again, pouring carefully over a spoon so that the two liquids do not mix. An egg placed in the jar remains suspended in the middle.

Since the egg is heavier than tap water, but lighter than salt water, it sinks only to the middle of the jar and floats on the salt water. You can use a raw potato instead of the egg. Cut a roundish "magic fish" from it, and make fins and eyes from colored cellophane.

134. Pearl diver

Stick a match about one-tenth of an inch deep into a colored plastic bead and shorten it so that its end floats exactly on the surface of the water when the bead is placed in a milk or juice bottle full of water. Close this with a plastic cap. Changing the pressure on the cap by pressing down on it, the bead can be made to rise and fall as though by magic. Plastic is only a little heavier than water. The match and the air in the hole of the bead give it just enough buoyancy for it to float. The pressure of the finger is conducted through the water and compresses the air in the bead. Thus it no longer has sufficient buoyancy, and sinks.

135. The yellow submarine

Cut a small submarine out of fresh lemon peel and make portholes on it with a ballpoint pen. Place the submarine in a bottle filled with water and close it with a plastic cap. If you press on the cap, the submarine rises and falls according to the strength of the pressure. Minute air bubbles in the porous fruit peel make it float. By the pressure of your finger, which is transmitted through the water, the bubbles are slightly compressed, so their buoyancy is reduced, and the submarine dives. Since the yellow of the peel is heavier than the white, the submarine floats horizontally.

You can accompany the submarine by several "frogmen." Simply toss broken-off match heads with it into the bottle. They float, because air is also contained in their porous structure. If the air bubbles are made smaller by the transmitted water pressure, the match heads dive deeper too.

136. Volcano underwater

Fill a small bottle full of hot water and color it with ink.
Lower the bottle by means of a string into a larger jar
containing cold water. A colored cloud, which spreads to
the surface of the water, rises upwards out of the small
bottle like a volcano.

Hot water occupies a greater volume than cold because
the space between the water molecules is increased on
heating. It is, therefore, lighter and buoyant in the cold
water. This heat-driven circulation is called convection.
After some time the warm and cold water mix and the ink
is evenly distributed.

137. Turning by buoyancy

A peach, placed in a glass of seltzer water, turns for quite a while, like a wheel. How does this work?

After you've filled the glass, the carbon dioxide, which has dissolved under pressure in the liquid, escapes. The gas gathers in little bubbles on the velvet-like peach skin, because the liquid has a large contact surface on its little hairs. As soon as one half of the peach has more bubbles and a larger buoyancy, it turns upwards. Since the bubbles burst in the air and new ones form on the bottom, the heavier portion remains on one side, causing the peach to slowly spin. A mill wheel works is a similar fashion. But while the wheel is moved by the weight of the water, the rolling peach is turned by the buoyancy of the gas bubbles.

138. Finding the center of gravity

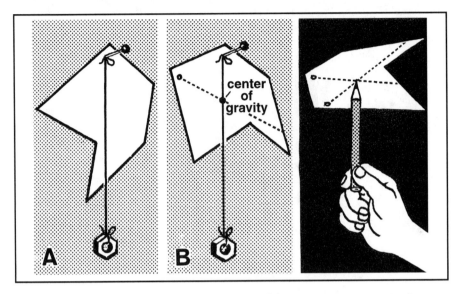

Anyone who wants to balance an object must support it under its center of gravity—the point around which all its mass is in balance. The center of gravity of an evenly shaped body made of a homogeneous material lies in its geometrical center. But how do you find the center of gravity of an unevenly shaped object?

It is simple to do for a piece of cardboard. Prick a needle through one corner and let the figure hang freely down. Its center of gravity must be somewhere directly below the suspension point. Draw a line as indicated by a thread hung onto the needle (figure A). Then hang up the cardboard from another corner (figure B) and draw another line. The center of gravity lies on the intersection of the two lines. Try to balance it on a pencil tip.

139. Center of gravity in trees

A free-standing fir tree resists a storm better than a fir tree standing in a forest. Why is that?

Since a free-standing tree receives sunlight, which is necessary for its growth, from all sides, its trunk, branches, and needles develop normally. In a forest, the light reaches the tree only from above. As these trees grow higher, their trunks become slender and breakable. Since their lower branches die, due to the lack of light, their center of gravity (S) lies higher in comparison to free-standing trees; their roots, which are spread laterally in shallow soil surface, provide a relatively unstable footing, almost as unstable as the clown which is placed on its head. Normally, the forest protects individual trees from the wind. But when part of it is cut, the tall, unstable trees will topple easily in a very strong wind.

GIANT BOOK OF SCIENCE EXPERIMENTS

140. Magic box

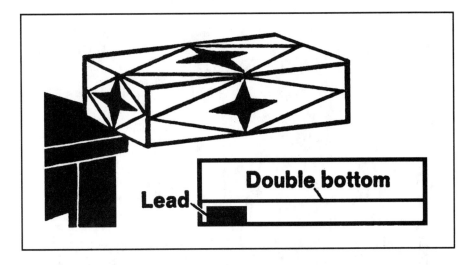

Lead

Double bottom

Create a false bottom in a thin cardboard box and hide a lead weight in the space below. You can always balance the box on the corner in which the piece of lead is lying.

Every object has a center of gravity. In every direction from this point, the downward pull of gravity is equal so that the object can be balanced. In such a regularly shaped object as a box the center of gravity is exactly in the middle, so your box should fall from the table. The lead weight prevents this, however, by shifting the center of gravity over the table.

141. Shifted center of gravity

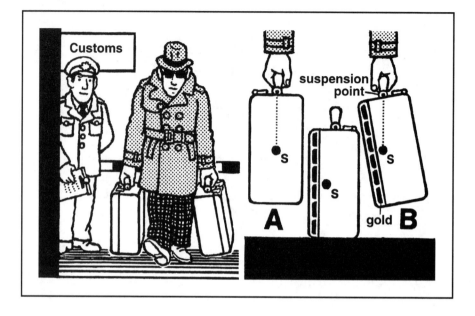

A customs officer observes a traveler whose suitcases (which he has inspected) appeared to contain only clothes. He becomes suspicious, and inspects the suitcases, finding a false bottom in one of them that contains bars of gold. What gives this suitcase away?

If a suitcase contains things which are all roughly the same weight, it hangs vertically, since it is an evenly shaped body and thus its center of gravity (S) is exactly in its middle (A). Due to the weight of the hidden gold, the center of gravity moves to the side (B). The suitcase tilts until its center of gravity is exactly beneath its suspension point (the handle).

142. The balancing button

If you place a button on a cup so that only the edges are in contact, it will fall off at once. No one would think that the button would remain on the rim of the cup if you placed yet another weight on it. And yet it is possible. If you fix two kitchen forks on the button and then place it on the rim of the cup, you can adjust them so that the button will remain in this position.

The bent fork handles, whose ends are particularly heavy and reach sideways around the cup, move the center of gravity of the button exactly over the rim of the cup, so that the whole setup is in balance.

143. Floating beam

Cut off

It would not seem possible to balance a clothespin with one end on the tip of your finger if a leather belt is hung over half the pin. But you can overcome the force of gravity.

The whole secret is a small groove which you cut slantwise in the piece of wood. The belt, which you squeeze firmly into the groove, leans towards you, shifting the center of gravity of wood and belt under the tip of your finger so that it balances.

144. Candle seesaw

Push a darning needle sideways through a cork and fix equal-sized candles at both ends. Then push a knitting needle lengthways through the cork and balance it over two glasses placed on top of a piece of aluminum foil. If you light the candles, they begin to swing.

Before the candles are lit, the center of gravity of the seesaw lies exactly on the axis so that both ends are balanced. But as soon as a drop of wax falls at one end, the center of gravity shifts to the other side. This is now heavier and swings down. Since the candles drip alternately, the center of gravity moves from one side to the other.

145. Letter scale

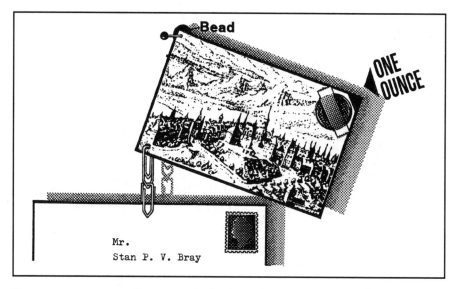

Tape a penny at the top right-hand corner of a picture postcard and fix two paper clips on the opposite corner. Hang the card up on a wooden surface with an easily turned pin in the top left-hand corner. The most simple of letter scales is made, and with it you can check the weight of a letter as accurately as with a normal type of letter scale.

You must first standardize your scale. Hang a letter that weighs exactly one ounce on the paper clips and mark the displacement of the top right-hand corner by an arrow on the wall. The scale moves farther for letters that weigh more than an ounce. This construction is a first-order lever, which is suspended by a pivot like a normal letter balance. The left-hand narrow edge of the card forms the loading arm. The upper edge forms the force arm, which shows, because it is longer, even small weight differences.

146. Braces for stability

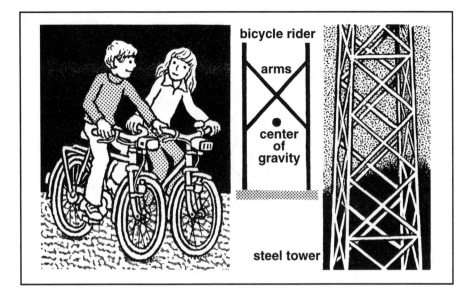

Two bicycle riders can stop, without falling over. They place the wheels of their bikes next to each other a few feet apart and each grabs, with one hand, the handlebars of the other bike. When the arms, which cross over, are straightened, the bikes stand firm. Why?

Their arms brace the two bikers in such a way above their common center of gravity that they can't tilt and fall over. Braces combined with vertical and horizontal construction elements form triangles which are very stable and are used widely in scaffolding and bridge construction.

147. Magic rod

Lay a rod over your index fingers so that one end sticks out farther than the other. Will the longer end become unbalanced if you move your finger farther towards the middle?

The rod remains balanced however much you move your finger. If one end becomes overweight, it presses more strongly on the finger concerned. The less loaded finger can now move farther along until the balance is restored. The process can be repeated by the combined effects of the force of gravity and friction until your fingers are exactly under the center of the rod.

148. Mysterious stability

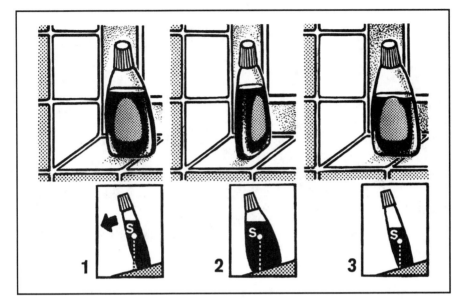

A shampoo bottle is placed on a slanted windowsill, first parallel to the window, then transversely to it, and after some shampoo is used to wash one's hair, again parallel. In which position does it tip over?

The stability of the bottle depends on whether its center of gravity (S) is supported by the surface it stands on. That is the case in position 2; the bottle is stable. In position 1, the perpendicular line of the center of gravity does not meet the standing surface, and the bottle tips over. In position 3, the center of gravity is shifted lower, after some shampoo has been taken out. But it remains unstable; since the perpendicular line of the center of gravity ends exactly at the edge of the bottle, a small push is sufficient to tip the bottle over.

149. Boomerang tin

Make slits half an inch wide in the middle of the bottom and lid of a round biscuit or cookie tin. Push a piece of thick rubber (a piece from a bicycle tire inner tube will do) the same length as the tin through the slits, and tighten it from the outside with pins. Hang a nut of about two ounces on the center of the rubber with a paper clip. If you roll the tin several yards forward, it will return at once.

The force of gravity prevents the nut from spinning with the tin as it rolls. It hangs down under the rubber and winds tighter with each rotation. When the force of the tension built up in the rubber is released, it causes the tin to roll back.

150. Balancing acrobat

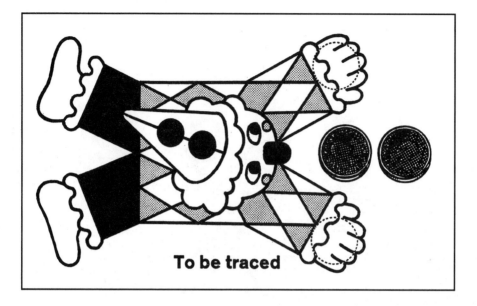

To be traced

Trace the large clown onto thick writing paper, cut out two figures and glue both pieces together with two small coins placed in the hands so that they are invisible. Color the figure brightly. The little paper clown will balance everywhere—on a pencil point, on your finger, or as a tightrope walker on a thread. This trick baffles everybody. It would seem that the figure should fall because its top half appears heavier.

The weight of the coins causes the center of gravity of the figure to shift under the nose, so that it remains balanced (see next page).

151. Principle of the aerial tram

After riding to the top of a mountain on an aerial tram, a father hikes down with his children toward the tram station at the bottom of the mountain. When they take a break along the path, the father sees the gondolas pass each other, and remarks, "We are now exactly halfway down the mountain!" How does he know that?

The two gondolas of the aerial tram hang on a supporting cable. They move up and down the mountain on a circular cable that is driven by the electrical motors in the uppermost station. The gondola that moves downwards counterbalances the one that moves upwards. Consequently, the gondolas depart on the top and bottom at the same time and meet exactly at half the way.

GRAVITY 163

152. Turning mechanism of an egg

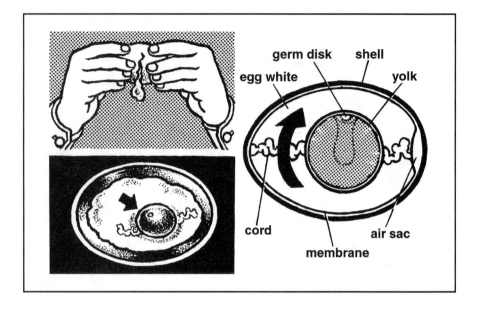

germ disk shell
egg white yolk
cord air sac
membrane

On the yolk of an egg there is a small white spot, the germ disk. Why does it always lie on top of the yolk in an egg cooked sunny-side up?

Before cracking it open, you can turn the egg whichever direction you like, around its longitudinal axis, but the yolk inside always turns with the germ disk upwards. This is because the yolk is suspended by a spiral-shaped cord in the egg white and its center of gravity is in the lower half. The germ disk, out of which the chicken develops during the process of brooding (the rest of the egg stores food and water for growth), is always on top, facing the warmth of the hen.

GIANT BOOK OF SCIENCE EXPERIMENTS

153. Paper bridge

Lay a sheet of writing paper as a bridge across two glasses, and place a third glass on it. The bridge collapses. But if you fold the paper as shown, it supports the weight of the glass.

Vertical surfaces are much less sensitive to pressure and stress than those laid flat. The load of the tumbler is distributed over several sloping paper walls. They are supported in the folds and thus have a very high stability. In practice the stability is increased by molding sheets and slabs to give rounded or angled sections. Think about the strength of corrugated iron and corrugated cardboard.

154. Unbreakable box

Put the outside case of a match box on a table, and on its striking surface place the box itself. Almost everybody will guess that the match box could be smashed with one blow of the fist. Try it! The box nearly always flies off undamaged in a high curve.

The match box is so strong because of its vertically joined sides that the pressure of the striking fist is transmitted to the outside without smashing it. The case, whose side walls seldom stand absolutely vertical, diverts the pressure to the side.

155. Cracks in the pipe

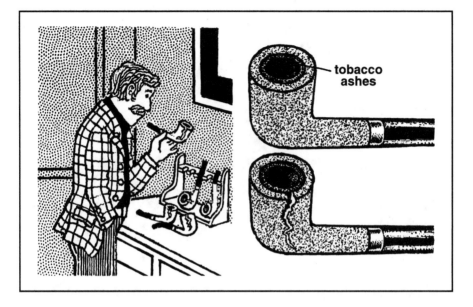

tobacco ashes

A man who became a non-smoker some months ago realizes, with surprise, that his tobacco pipes have cracked. How did this happen? When tobacco was burned in them, the humidity that was contained in the little leaves turned into water vapor, entered the pores of the pipe wood, and caused it to swell in the course of time. At the same time, tobacco ash was deposited on the inner wall of the pipe, and soon formed a thick layer. When the pipe was not smoked anymore, the wood dried out and shrunk. But since the stone-hard ash layer kept its shape, the tension that was created caused the wood to crack.

156. Strong egg

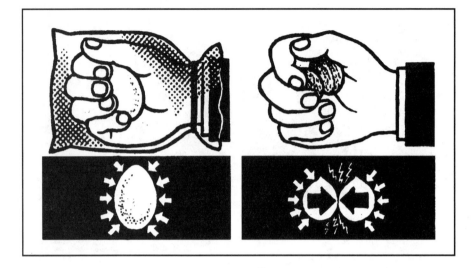

Who wants to bet that you can crack walnuts more easily in your hand than a raw hen's egg? Take a slightly polished egg, place your hand as a precaution in a plastic bag, and squeeze as hard as you can!

The lever pressure of the fingers is distributed evenly from all sides onto the egg and is not enough—if the shell is undamaged—to break it. Curved surfaces are extremely strong. People use this advantage in the building of arches and bridges, and cars hardly have a flat surface for the same reason. Two walnuts, however, can be easily cracked in one hand, because the pressure is concentrated at the points of contact.

157. Visible molecular force

In a box of fancy chocolates you can find a transparent cover sheet made out of cellophane, a material similar to cellulose. It serves a very specific purpose: the sheet regulates the humidity in the package. Cellophane is made from wood, and is hygroscopical, that is, it absorbs water vapor from the air by expanding.

Experiment: Take a strip of cellophane about 4 inches long and clamp it between your fingers so that with your hand flat, it flips constantly from one side to the other. The side that is turned towards the hand expands as it absorbs water molecules which are escaping from the pores of the skin. This expansion on one side causes the paper to flip over. After turning over, the paper passes the water on to the dry air and contracts again.

158. Spinning ball

Place a marble on the table, with a jam jar upside down over it. You can lift the ball in the jar and carry it as far as you like, without turning it the right way up. Anybody who does not think it is possible should make a bet with you.

This trick is made possible by making circular movements with the jar which set the ball rotating. The ball is pressed against the inner wall of the jar by centrifugal force. The narrowing of the glass jar at its mouth prevents the ball from dropping out when you lift it from the table.

GIANT BOOK OF SCIENCE EXPERIMENTS

159. Egg-top

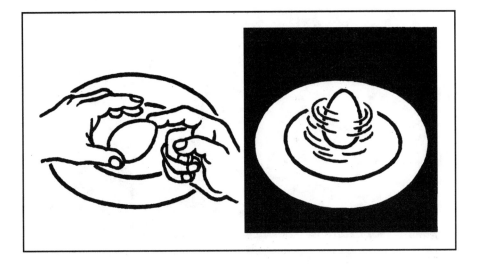

There is a very simple method for distinguishing a cooked egg from a raw one without breaking the shell. Spin the eggs on a plate, and the well-cooked one will continue to rotate. Since its center of gravity lies in the thicker half, it even stays upright, like a top.

The liquid inside the raw egg prevents this. Since the yolk is heavier than the white, it rolls from the middle when you spin the egg, because of centrifugal force. It brakes the movement so much that it only amounts to a clumsy rocking.

160. Coin bumping

Lay two coins in a row on a table so that they touch. Press one hard with your thumb and flick a third coin against it. The neighboring coin shoots away, although the middle one is held firm.

Solid bodies possess a large elasticity that is shown, for example, in steel when it is made into a spring. In our experiment the coins are imperceptibly compressed when they collide but spring back at once to their original shape and transfer the impact to the neighboring coin in this way.

161. Dividing money

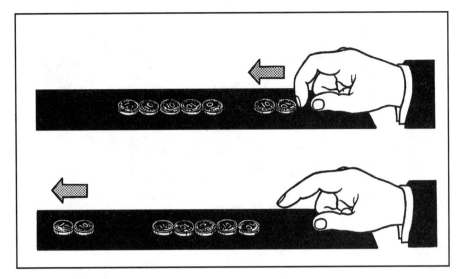

Lay several coins of the same value in a row on an even surface so that they touch. Lay another coin a short distance away from the row, and flick it with your finger against the others. What happens? A coin slides away at the other end of the row. Repeat the experiment, propelling two coins against the row. This time two coins separate off. If you flick three coins, three are removed, and so on. In a further experiment you can flick the coins really hard, but with no different result.

The result is truly surprising, but illustrates a physical law. The coins undergo an elastic impact when they are knocked together, and the same weight (number of coins) as the striking coins carries on the movement at the other end of the row. The sharpness of the flip decides how fast and how far the coins fly off, but has no effect on their number.

162. Uncuttable paper

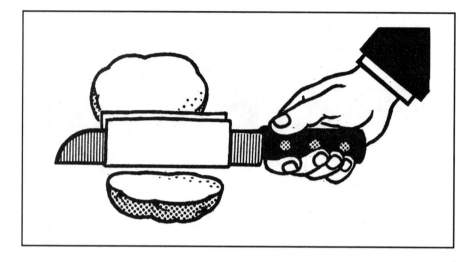

Place a folded piece of writing paper around a knife blade. You can cut potatoes with it without damaging the paper.

The paper is forced into the potato with the knife. It is not cut itself because the pressure of the blade on the paper is countered by the resistance from the potato. Since its flesh is softer than the paper fiber, it yields. If, however, you hold the paper firmly on top, the pressure balance is lost, and the paper is broken.

163. Increasing friction resistance

After a boy has watered one half of a lawn, he pulls the hose to water the other side of the garden. Why is it that at first the boy finds the hose to be very easy to pull, but with each step he has to use more force to drag it until he can barely move the hose?

The farther the boy walks, the longer the section of the hose behind the bend which he has to move becomes. Since the hose is filled with water, its weight increases with each foot added. The resistance of friction, which the boy has to overcome, grows with the weight of the hose dragged across the lawn.

164. Effect of a thousand levers

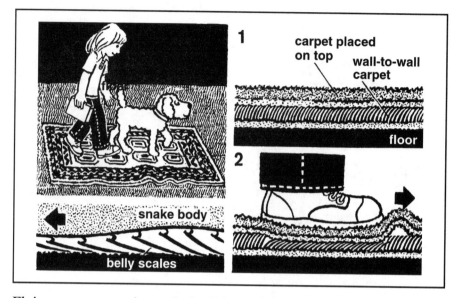

1 carpet placed on top wall-to-wall carpet floor

2

snake body

belly scales

Flying carpets exist only in fairy tales, but wandering carpets are real. If you walk over a carpet that is placed on a wall-to-wall carpet, it changes its position.

The fibers of the wall-to-wall-carpet, which, through a special kind of weaving, stand all in one direction, rest against the underside of the carpet placed over it. When you step on the carpet, the fibers of the wall-to-wall carpet bend to the side. Each fiber acts like a lever, and the force of all the fibers together pushes, with every step, the respective part of the carpet forward the length of fiber.

Snakes crawl in a similar way. Their wide belly scales are angled one after another from the front backwards and are laid against their bellies again. The scales catch any unevenness of the terrain and push the animal's body forward.

165. Problem with a cart

Pushing a cart up a step is difficult. The rubber wheel gets caught tight in the 90° angle. But when you reverse the cart and pull it backwards, the wheel easily rolls up over the step. This is easy to understand when you consider the spot (D) where the wheel touches the step on the vertical surface. In case 1, the force (K), transmitted from the cart handles, is directed below (D) and causes the wheel to turn slightly backwards. It therefore becomes stuck in the angle of the stair. In case 2, the force (K) is directed above (D) so that the wheel rolls up the stair.

166. Turning a spool

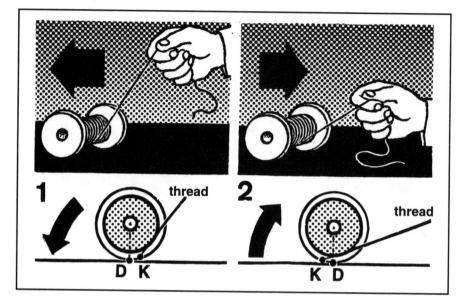

A spool of thread rolls away, leaving a piece of the thread unwound. Can you retrieve the spool by pulling on the thread? The answer depends on the angle at which you pull: if you pull the thread steeply upwards, the spool rolls away; but if you pull it flatly towards you, it rolls the spool in the same direction.

The different turning directions are caused by the fact that the turning axle (D) is not—as you might think—in the center of the spool, but where it touches the floor. The force (K) on the spool is in the direction of the thread and moves it around the turning point. In case 1, force (K) acts on the spool in front of the turning point (D) and turns the spool backwards. In case 2, it acts on the spool behind (D) and causes a forward movement.

GIANT BOOK OF SCIENCE EXPERIMENTS

167. Forces on a toboggan

A boy pulls a toboggan sled on an icy surface. After a piece of the cord of the sled has broken off, the boy thinks that it has become harder to pull the sled than it was before. Is he right?

A diagram of the forces acting on the sled clarify how the length of the cord influences the movement of the sled. With a long cord (A), the boy pulls with the force (K) slightly slantingly upwards. This force can be visualized as two separate forces, K_1 and K_2. K_1 is important because it alone moves the sled forward, while K_2 unnecessarily lifts upward the sled. In the case of the shortened cord (B), the relationship of the forces is worse; K_2 remains the same, but K_1 has decreased. With an even shorter cord (C), the sled would be more lifted than pulled—as one does it, if need be, in high, wet snow.

168. Distribution of forces

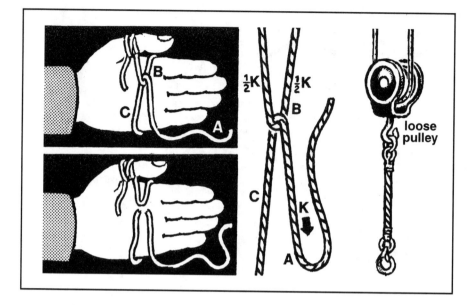

It is surprisingly easy to break a cord, when you place it, as in the illustration, around your hand and pull sharply at the end (A). This end always breaks off above the loop B. Why?

During the jerky pull, the force (K) transmits itself mainly to the loop (B), leaving section (C) relatively unaffected. The force distributes itself onto the two halves of the loop, which fit tightly around the thumb and the back of the hand so that only half of the force acts on each of them. Therefore, the end of the cord (A) is stressed most, and its fibers are weakened by the friction above the loop. The distribution of forces on a loose pulley of a crane is similar. The cord, which leads around the pulley, needs to be only half as strong as the cord, which hangs below it.

169. Inertia on the train

An empty bottle lies on the floor of a train compartment. It rolls back and forth each time the train stops and starts. What forces are responsible for moving it? According to Newton's first law, a body remains in a state of rest or constant velocity unless a force acts upon it. This resistance to change its state of movement is called inertia. When the train starts to move, the bottle wants to remain in its state of rest, but the floor of the car literally pulls itself away underneath the bottle. When the train brakes, the bottle wants to continue moving and rolls into the direction the train was headed. When the train moves at a steady speed, the bottle lies motionless.

170. Proof of inertia

A floor tiler is stopped in his van by the police, who accuse him of having just caused an accident involving the cars behind him. They claim that at a crossing he suddenly hit his brakes. What proof is there that the driver drove carefully and braked gently?

The piles of tiles in the back of the van are standing upright. If the driver had braked suddenly, the tiles (which, according to their weight, have a large inertia and which easily slide on their smooth surfaces due to low friction) would have flown forward in the direction of motion.

171. The stable pencil

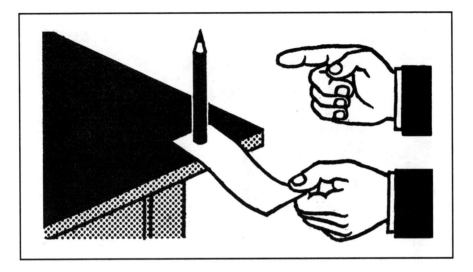

Hold a strip of paper over a smooth table edge and place a pencil on it. Can you remove the paper without touching the pencil or knocking it over? The pencil will certainly fall if you pull the paper away slowly. The experiment works if you jerk the paper away in an instant by hitting it with your finger.

Each body tries to remain in the position or state of motion in which it finds itself. The pencil resists the rapid movement, so that it remains where it is and does not tip over.

172. Treasure in the tower

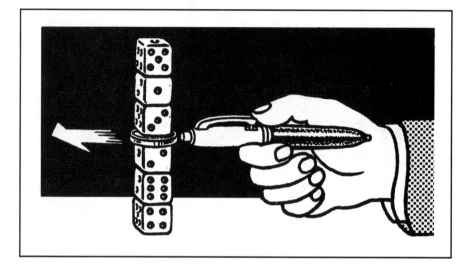

Build six dice into a tower and place a quarter in the middle. The tower is very rickety, so how can you remove the coin without touching it or knocking it over?

Hold a ballpoint pen with a pushing clip a little distance from the coin. If you discharge it the coin flies out of the tower. The movement of the spiral spring in the ballpoint pen is transferred at once to the coin, but, because of the low friction, not to the dice, which because of their weight have a fairly large inertia.

173. Egg bomb

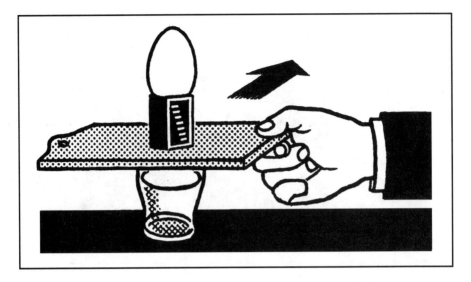

Lay a small board on a glass of water, place a matchbox case on it and on this a raw egg. Can you transfer the egg to the water without touching it? Pull the board sharply to the side! The egg falls undamaged into the water.

Because of its weight, the inertia of the egg is so great that it is not carried along with the fast movement. The light matchbox, on the other hand, flies off because its inertia is low.

174. The lazy log

Tie two pieces of string of equal thickness to a block of wood or another heavy object. Hang the wood up by one string and pull on the other. Which section of the string will break?

   If you pull slowly, the strain and the additional weight of the object causes the upper string to break. But if you pull jerkily, the inertia of the block prevents the transfer of the total force to the upper string, and the lower one breaks.

GIANT BOOK OF SCIENCE EXPERIMENTS

175. Dividing an apple

Cut far enough into the flesh of an apple with a knife so that when you lift it up it sticks on the blade. Now knock against the blade with the back of another knife. After several blows the apple will be cut in half.

In the sixteenth century, the famous Italian scientist Galileo discovered that all bodies resist change to their position or state of motion, which we call inertia. This prevents the apple from following the blows of the knife. It pushes slowly on the blade, until it is cut.

176. Coin shooting

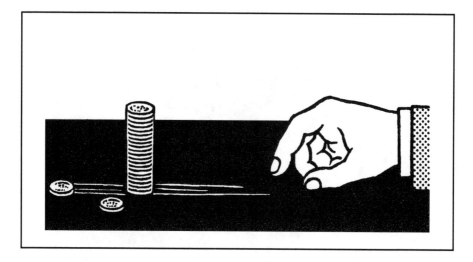

Place about twenty identical coins in a pile on a smooth table. How can you take away the coins one at a time from below, without touching them? Flick another coin sharply with your finger so that it hits the bottom coin and shoots it away.

If you aim well, you can shoot away all the coins in this way. The inertia of the coin column is so great that the force of the flicked coin is not sufficient to move it or overturn it.

177. Inertia of the gases

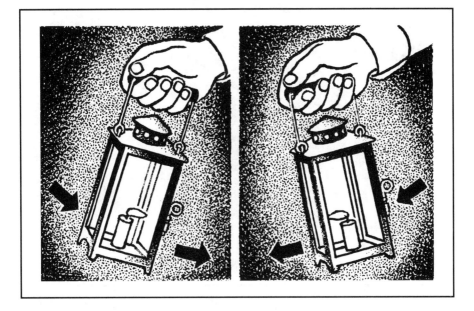

When you swing a candle lamp back and forth, you realize that the flame moves from one side to the other. But strangely it moves in the direction of the motion and not, as you might expect, in the opposite direction. Why is that?

Since the air in the lantern is cooler and thus heavier than the hot gas in the flame, the air also has a larger inertia. The air presses against the side of the lantern opposite to the direction of motion and creates a slightly higher pressure there. Each time, the flame turns to the side which is thinned by air. Similarly, water accumulates at the edge of a glass or bucket when it is swung.

178. Centrifugal force

A riding ring, in which the horses run in a circle, becomes more and more inclined in the course of time. Like the edge of a plate, it becomes considerably higher on the outside than on the inside. Why is that?

When running in a circle, the inertia, which is directed from the center of the ring towards the outside, affects the horse and rider. This can be seen as little pieces of dirt that fall off the hoofs of the horses. They do not follow the circular movement of the horse, but instead they fly a little bit to the outside in the direction of the arrows tangent to the circle.

The same inertia is visible when you stir coffee in a cup. The grains, if any, move towards the edge and accumulate there.

179. Variable wheel circumference

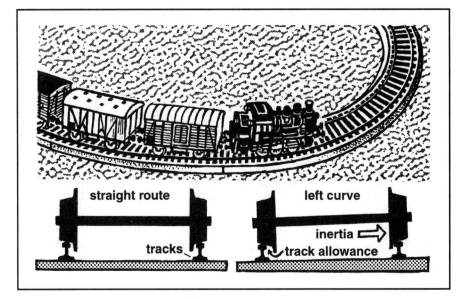

straight route   **left curve**

tracks   inertia ⟹   track allowance

On the circular track of a toy train that is 72 inches long, the outer rail is 13 inches longer than the inner one. On the outer rail, therefore, the wheels of the train ought to turn faster than on the inner one; but they are firmly attached to the same axle. How is this problem solved?

The wheels have (just like the ones of large trains) some allowance to shift sideways on the track on conical wheel treads. Thus they roll with different diameters against the rails when shifted laterally. On a straight track, this keeps the train from wandering off center. In a curve, inertia shifts the wheels to the outside of the curve so that they roll on the outer rail on a larger wheel circumference, which covers more distance per revolution, than on the wheel on the inside track.

180. Action and reaction

Some people believe that you can make yourself lighter on a scale by slowly bending your knees. Is this correct, and if so, does your indicated weight increase when you stretch your arms upwards?

According to Newton's third law, to every action there is an equal and opposite reaction. So, throwing your arms up forces your body downwards, exerting more pressure on the scale, and the indicator moves to the right. When the arms are fully stretched and their inertia pulls upwards, the pressure onto the scale decreases and the indicator briefly turns to the left. When you bend your knees, the opposite happens and the scale indicates less weight. Only when the movement is finished does the scale indicate more weight.

GIANT BOOK OF SCIENCE EXPERIMENTS

181. Opposite forces

A freight train is parked next to a loading platform, while a heavily loaded truck rolls along on top of it. Why does the train have to be fastened to the train tracks to keep it from moving?

According to Newton's third law, without being fastened, the train would slowly roll backwards underneath the moving truck.

Experiment: If you place a ruler over two round pencils and put a wind-up toy on top of it, the toy moves forward, while the ruler moves backwards. If both objects have the same weight, they will have the same speed. If the ruler is heavier than the toy, the toy will roll faster. If the ruler is lighter than the toy, the ruler rolls faster.

182. Humming flute

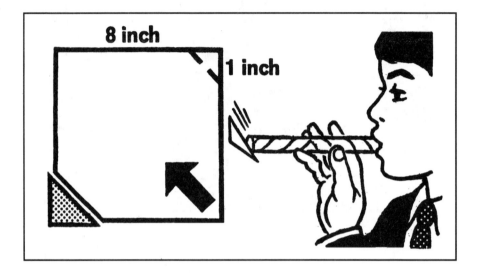

Take a square piece of paper and snip one corner off, then make two notches in the opposite corner. Roll the paper in the direction of the arrow in the figure to make a tube about as thick as a pencil, and fold the notched corner to cover the opening. Draw a deep breath through the tube. This causes a loud humming sound. The paper corner is sucked up by the air drawn in, but since it is slightly springy, it begins to vibrate. The vibration is quite slow, so it makes a deep note.

183. Musical drinking straw

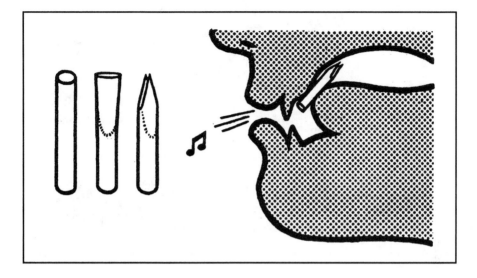

Cut a piece about an inch long from a plastic drinking straw. Press one end together and cut it to a point. If you place the straw against the front of your top palate, you can make various notes when you blow through it.

The pointed tongues of the straw move rapidly as the air sweeps through, so that the note produced is fairly high. A great number of musical instruments are based on this simple principle.

184. Water organ

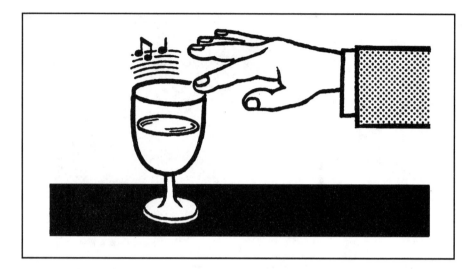

Half fill a thin-walled glass with water, dip in your forefinger, and run it slowly around the rim of the glass. With a little practice, a lovely, continuous ringing note is produced.

This experiment works best if you have just washed your finger. It rubs over the glass, giving it tiny jolts. The glass begins to vibrate, which produces the note. (The vibration can be seen clearly on the water surface.) If your finger is greasy, it slides smoothly over the glass without the necessary friction. The pitch of the note depends on the amount of water in the glass.

185. Note transfer

You can extend the previous experiment. Place two similar thin walled glasses an inch part and pass your freshly washed finger slowly round the rim of one of them. A loud humming note is produced. Mysteriously, the second glass vibrates with the first. You can observe this vibration if you place a thin wire across the second glass.

The vibration of the first glass is transmitted to the second by the sound waves in the air. This "resonance" only occurs if the glasses produce notes of the same pitch when struck. If this is not the case with your two glasses, you must pour some water into one until the pitches are the same.

186. Peal of bells

Tie a fork in the middle of a piece of string about a yard long. Wind the ends several times around your forefingers and hold the tips of your fingers in your ears. Let the fork strike a hard object. If the string is then stretched, you will hear a loud, bell-like peal.

The metal vibrates like a tuning fork when it strikes the hard object. The vibration is not carried through the air in this case, but through the string, and the finger conducts it directly to the eardrum.

187. Footsteps in a bag

Trap a housefly in a smooth paper bag, seal it, and hold it horizontally above your ear. If you are in a quiet room you can hear the patter of the six legs and other rather curious noises quite clearly.

The paper behaves like the skin of a drum. Although only the tiny legs of the insect beat on it, it begins to vibrate and transmits such a loud noise that you might imagine a much larger animal was in the bag.

188. Box horn

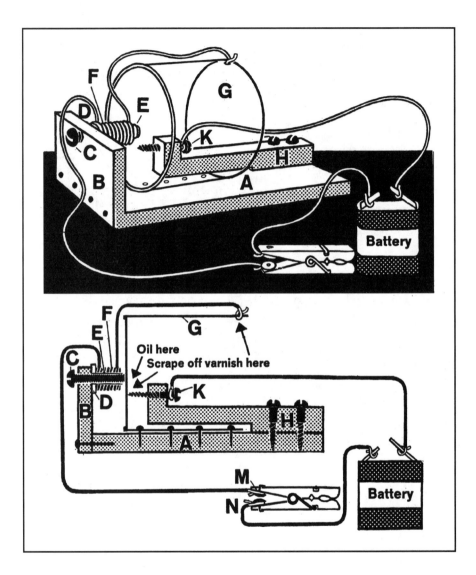

Nail as large a tin can as possible firmly onto a suitable board A. Through board B, which is nailed at the side, bore a hole by which you can turn an iron screw C to the middle of the base of the can. A small space should remain between the screw and the can. Put a layer of paper E around the screw, and over it wind about 2 yards of insulated wire F. From the inside, a wood screw K that is fixed so it can be moved in a piece of wood H contacts the base of the can. Scrape off the metal plating in front of the tip of the screw and oil it. Join all the parts correctly with connecting wire, removing the insulation and tin varnish from the connecting points. A clothespin with two metal thumbtacks M and N acts as the horn switch. If you press it, you will hear a very loud noise.

The apparatus works on the same principle as a car horn. If you close the circuit by pressing the horn switch, the screw C becomes magnetic and attracts the base of the can. So the circuit is broken in front of the screw K. Screw C loses its magnetism, and the base of the can springs back to the screw K. The process is repeated so quickly that the tin plate produces the horn blast by its vibration.

189. Speed of sound

Since sound waves travel through the air at a rate of roughly one mile every 5 seconds, you can calculate the distance in miles away of a lightning bolt by dividing the number of seconds between the lightning flash and the thunder by 5. But why does the rumble of the thunder last so much longer than the lightning flash?

Thunder is caused by the rapid expansion of air that is heated to very high temperatures along the entire length of the lightning bolt, which may often be a mile or more long. The sound waves from this "long" explosion need different amounts of time to reach our ears, because some parts of the lightning bolt are farther away from us than others. After you hear the bang, which is delayed and weakened with increasing distance, you can often hear a weak rumbling sound, which is the reflected sound waves.

GIANT BOOK OF SCIENCE EXPERIMENTS

190. Pinhole camera

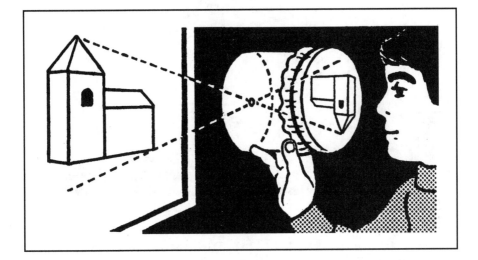

Poke a small hole in the middle of the base of a round box. Stretch translucent parchment paper (or wax paper) over the mouth of the box and secure it with a rubber band. If you focus this simple camera on a brightly lit building from a dark room, the image appears upside down on the screen.

Our eyes work on the same principle. The light rays fall through the pupil and lens and project an inverted image on the retina. Because our brains process the image our retina receives, the image is perceived the right way up.

191. Drop microscope

Punch a hole about one fifth of an inch wide in a strip of metal and smooth the edges. Bend the metal so that you can attach it with adhesive tape half an inch above the bottom of a thin glass turned upside down. A pocket mirror is placed inside on a cork, so that it slants. If you dab a drop of water into the hole, you can see small living organisms and other things through it, magnified by up to fifty times.

The drop magnifies like a convex lens. When you bring your eye near it the sharpness can be adjusted by bending the metal inwards. The angle of the mirror is adjusted automatically by moving the glass to provide light for the microscope.

192. Fire through ice

You would hardly believe it, but you can light a fire with ice! Pour some water that you have previously boiled for several minutes into a symmetrically curved bowl, and freeze it. You can remove the ice by heating it slightly in warm water. You can concentrate the sun's rays with the ice as you would with a magnifying glass and light, thin black paper. Black paper works best because it absorbs the heat rather than reflecting it.

The air in freshwater forms tiny bubbles on freezing that makes the ice cloudy. But boiled water contains hardly any air and freezes to give clear ice. The sun's rays are only cooled imperceptibly when they pass through the ice.

193. Shortened spoon

Look from just above the rim of a bucket of water, and dip a spoon upright into it. The spoon seems to be considerably shorter under the water.

This illusion is based on the fact that the light rays reflected from the immersed spoon do not travel in a straight line to your eyes. They are bent at an angle at the surface of the water, so that you see the end of the spoon higher up. Water always seems more shallow than it actually is, because of the refraction of light. The American Indians also knew this. If they wanted to hit a fish with an arrow or spear, they had to aim a good deal deeper than the spot where the fish appeared to be.

194. Shadow play

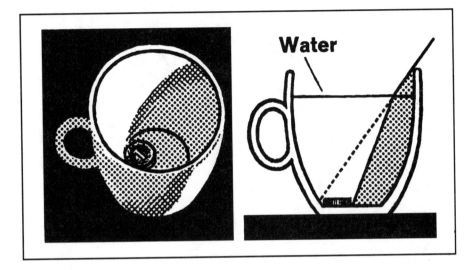

Water

Lay a penny in a cup near the side. Place the cup in oblique light so that the shadow of the rim just covers the coin. How can you free the penny from the shadow without moving the cup or the coin or using a pocket mirror?

Quite simple! Bend the light rays back to the coin. Fill the cup with water and the shadow moves to the side. The light rays do not go on in a straight line after striking the surface of the water, but are bent downwards at an angle.

195. Broken pencil

Fill a glass halfway with a concentrated salt solution and then slowly fill it to the top with pure water, using a spoon so the two layers won't mix. If you hold a pencil to the side in the glass, it seems to be broken into three pieces.

The first apparent break occurs because the light rays coming from the immersed pencil are bent at an angle when they emerge from the water into the air at the side of the glass. The second break occurs because the salt water has a different composition from pure water and the angle of refraction is different. How much light rays are bent when they pass from one substance into another depends entirely on the "optical density" of each substance.

196. Cloud of gas

If you pour some bicarbonate of soda (baking soda) and vinegar into a beaker, carbon dioxide is given off. You can normally not see the gas, but it can be made visible if you tilt the beaker with its foaming contents in front of a light background in sunlight. You can see the gas, which is heavier than air, flowing from the beaker in dark and light areas.

Carbon dioxide and air have different optical densities, and so the light rays are bent when they pass through them. The light areas on the wall are formed where by refraction the spread-out light is bent towards it, and the dark areas are seen where light is bent away.

197. Magic pencils

Look through a round jam jar filled with water. If you stand a pencil a foot behind it, its image appears doubled in the jar. If you close your left eye, the right-hand pencil disappears, and if you close your right eye, the other goes.

Through a normal magnifying glass, you see distant objects reduced in size. The water container behaves in a similar way, but since it is cylindrical, you can look through it from all directions. In our experiment both eyes look through the jar from a different angle, so that each one sees a smaller image for itself.

198. Invisible People

During a break on a path in the forest, the children in a car are surprised when a rabbit hops out of the preserve and grazes directly next to the car. "It can't see us in the car," the father says. Is he right?

When the rabbit sits so close to the car, it sees from below only the reflection of the sky and the trees on the car windows, which are slanted. It does not see the inside of the car and the people. Since it is darker in the car than outside, the image of the people is outshone by the bright image that is reflected on the windows. The light beams hit upon the glass surface in such a large angle of incidence that they are (in an angle of reflection just as large) almost completely reflected towards the rabbit's eyes.

199. Light trap

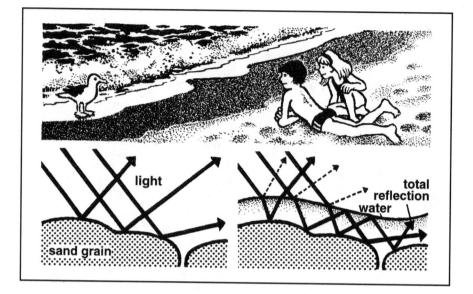

Wet sand appears darker than dry sand. If water is colorless and transparent, why is this so?

Sand consists of tiny grains of quartz. The smooth surfaces on each grain reflect the sun's rays in all directions. Therefore, dry sand looks almost white. In wet sand, a portion of the light is swallowed up by a thin layer of water which surrounds the each grain. Some of the light rays are reflected from the sand grains in such a way that they strike the surface between the air and water and are reflected back downwards into the water. This means that only the rays that strike at a steep angle where the air and water meet will reach the eye (in addition to the light which is already being reflected above the water layer and which causes the glitter).

200. Finger heater

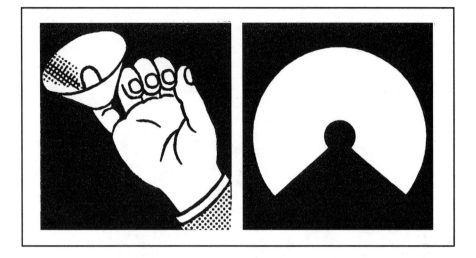

Glue a funnel together with smooth silver paper, as shown in the picture. Stick your finger into it, point it to the midday sun, and you will feel it warm up.

The sun's rays are reflected from the walls of the funnel to the middle and are concentrated on the central axis, where your finger lies. If you put your finger into the dismantled concave mirror of a bicycle light, the sun's rays would be unbearably hot. In this case they converge at a point, the focal point of the concave mirror, at which the bulb is usually placed. The heat produced is so great that one could easily start a fire with a concave mirror.

201. Sun power station

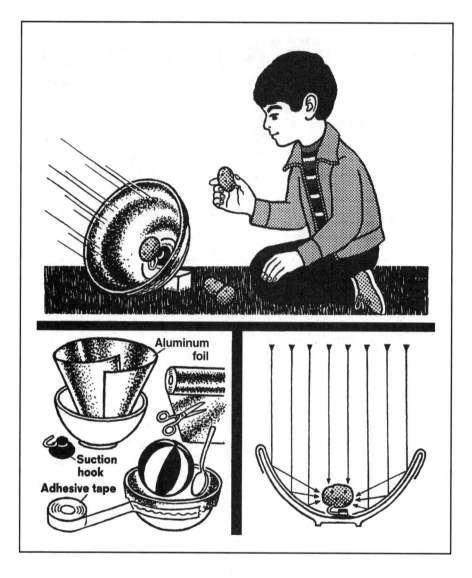

The sun's radiation can be concentrated in a bowl and used to cook. Take a soup bowl or a large salad bowl with as small a base as possible and line it inside with household aluminum foil—bright side outwards. Smooth the folds with a rubber ball and a spoon until the foil acts like a mirror. Remove a little foil at the base of the bowl so you can press in a suction cup with a hook, on which you attach a small raw potato. On a warm day if you point the cooker towards the midday sun, the potato becomes hot at once and is cooked after some time.

Because the earth is spinning, you must occasionally realign the bowl towards the sun. The sun's rays falling on the aluminum foil are reflected to the middle and concentrated on the potato. In tropical countries people often use concave mirrors for cooking. Did you know that electricity can be produced in large power stations by the sun's radiation?

202. Magic glass

If you place a jar over a coin lying on a table, it looks just as if it were in the jar. If you pour water into the jar and put the lid on it—abracadabra!—the coin has disappeared, as if it had dissolved in the water.

When the jar is empty the light rays reflecting off the coin travel into our eyes in the usual way. But if the jar is filled with water, the light rays no longer follow this path. They are reflected back from the bottom of the glass when they hit the water from below at an angle. We call this total reflection, and only a silvery gleam can be seen on the bottom of the jar.

203. View into infinity

Hold a pocket mirror between your eyes so that you can look to both sides into a larger mirror. If you place the mirrors parallel to one another, you will see an unending, infinite, series of mirrors which stretches into the distance like a glass canal.

Since the glass of the mirror shines with a slightly greenish tint, some light is absorbed at each reflection, so that the image becomes less sharp with increasing distance.

204. Mirror cabinet

Obtain three sections of mirror each about 3 × 4 inches in size. Clean them well, and join them with adhesive tape—reflecting surfaces facing inwards—to make a triangular tube. Stick colored paper outside. If you look obliquely from above into the mirror prism you will see a magic world of optical illusions. If you hold a finger in the prism, its image is always multiplied six times in an endless series in all directions. If you place a small flower inside, a meadow of flowers stretches into the distance. And if you move two small figures, innumerable couples dance in an immense hall of mirrors.

205. Shining head

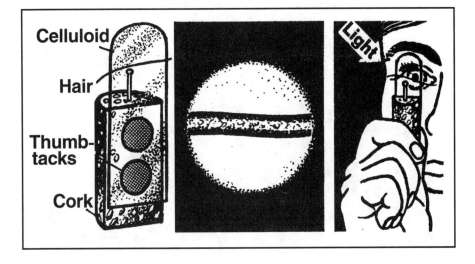

Celluloid
Hair
Thumb-tacks
Cork
Light

Stick a pin with a polished head into a cork cut in half lengthwise and attach some celluloid to it with thumbtacks to protect your eyes. If you look at the tiny light reflection from the head of the pin under a bright lamp, while holding it right up to the eye, it appears as a plate-sized circle of light. A hair stuck onto the moistened celluloid is seen magnified to the width of a finger in the circle of light.

The head of the pin behaves like a small convex mirror. The light which hits it is spread out on reflection, and irradiates a correspondingly large field on the retina of the eye.

206. Light mill

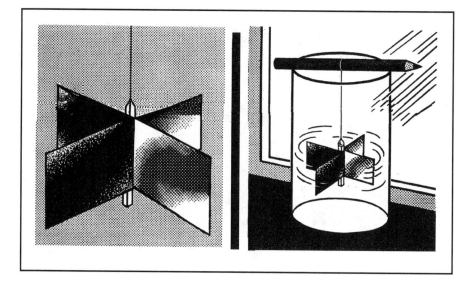

Cut out four pieces of aluminum foil $1 \times 1\frac{1}{2}$ inches in size. Glue the sheets onto a match like the blades of a mill wheel, with the bright sides all facing in the same direction. Blacken the matt sides over a candle, holding a knife blade behind the foil to assist you. Glue a hair, or fine thread, to one end of the match and let it dry. Place a tall jar in the sun, hang the mill inside, and it soon begins to turn without stopping.

We know that dark surfaces are more strongly heated by sunlight than light ones. This heat difference is the secret of the light mill. The sooty side of the foil absorbs the light rays and is heated about ten times more strongly than the light-reflecting bright side. The difference in the amount of heat radiated from the sides of the blades causes the rotation.

207. The sun's spectrum

Lay a piece of white paper on the windowsill and place on it a polished glass full to the brim with water. Tape a postcard with a finger-width slit onto the glass, so that a band of sunlight falls on the surface of the water. A splendid spectrum appears on the paper. Bands of red, orange, yellow, green, blue, indigo, and violet light can be easily distinguished. Each color light has its own wavelength, which determines how much it bends when it passes through the glass.

The experiment is only possible in the morning or evening, when the sunlight falls at an angle. It is refracted at the surface of the water and again at the side of the glass, and is separated as well into its colored components.

208. Spectrum in a feather

Hold a large bird's feather just in front of one eye and look at a burning candle standing a yard away. The flame seems to be multiplied in an X-shaped arrangement, and also shimmers in the spectral colors.

The pattern is produced as light passes between the fine vanes and barbs of the feather. The light is bent as it passes through them, that is, it is refracted and separated into the spectral colors. Since you see through several slits at the same time, the flame appears many times. These structures are regularly spaced and form narrow slits with sharp edges.

209. Colored hoop

Rainbows in the sky always appear as a semicircle. But you can create a complete circle for yourself from sunlight. Stand outdoors on a stool in the late afternoon with your back to the sun and spray a fine mist of water with the hose. A circular rainbow appears in front of you!

The sunlight is reflected in the water drops so that each shines with the spectral colors. But the colors of the drops are only visible to your eyes when they fall in a circular zone at a viewing angle of 85° in front of you. Only the shadow of your body briefly breaks the circle.

210. Reflection in ice crystals

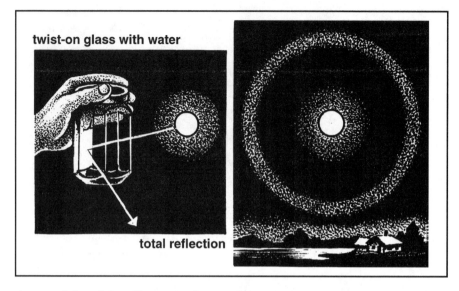

twist-on glass with water

total reflection

A considerable distance from the moon or sun, you occasionally can see a halo, or a large ring of light. What causes this phenomenon?

The light from the celestial body passes through a thin high-altitude cloud of hexagonal ice needles and is totally reflected by a surface of these crystals. But you only see the result of the countless little reflections in a ring-shaped zone of ice crystals, which is located at a specific angle to your eye. You can make the same observation when you look in the evening at a bright street lamp through a pane of glass which is slightly covered with white frost.

Experiment: Hold a hexagonal jar, filled with water, laterally in front of the moon, and you will see—just as in the ice crystal—the total reflection of the moonlight on one inside surface of the glass.

211. Colored top

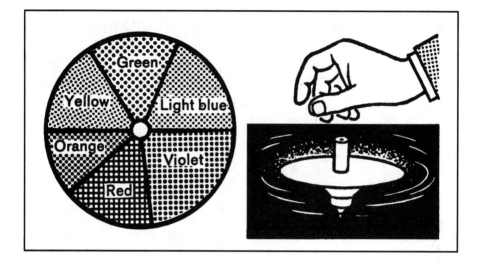

Cut a circle about four inches in diameter from white cardboard and color it as shown with bright-colored felt-tip pens. Glue the disk on a wooden thread spool cut in half, push a pencil stump through it, and allow it to spin. The colors disappear as if by magic, and the disk appears white.

The colors on the disk correspond to the colors of the spectrum of which sunlight is composed. When they are spun, our eyes perceive the individual colors for a very short time. However, since the eyes are too sluggish to distinguish between the rapidly changing color impressions, they merge and are transmitted to the brain as white.

212. Fiber optics

Through thin, flexible light-conducting fibers made of glass or plastic, you can send a beam of light and you can transfer images. Light-conducting fiber optics are also used in the communications industry as well as in medicine.

The beams of a light pass through a bent fiber in the zigzag path and are totally reflected off the sides over and over again.

Experiment: Stick a thin penlight (in a watertight transparent bag) inside the spout of a watering can. When you are watering in the dark, the arch of water glitters a little, because some light beams escape to the outside through its rippled surface. Where the water hits the ground, a spotlight can be seen.

213. Unusual magnification

Make a small hole in a card with a needle. Hold it close to your eye and look through it. If you bring a newspaper very close you will see, to your surprise, the type much larger and clearer.

This phenomenon is caused by the refraction of light. The light rays passing through the small hole are made to bend and spread out, and so the letters appear larger. The sharpness of the image is caused—as in a camera—by the shuttering effect of the small opening. The part of the light radiation which would make the image blurred is held back.

214. Veined figure

Close your left eye in a dark room and hold a flashlight close beside your right eye. Now look straight ahead and move the flashlight slowly to your forehead and back. After some time you will see a large, treelike branched image in front of you.

Very fine blood vessels lie over the retina of the eye, but we do not normally see them. If they are illuminated from the side, they throw shadows on the optic nerves lying below and give the impression of an image floating in front of you.

215. Motes in the eye

Make a hole in a card with a needle and look through it at a dim electric light bulb. You will see peculiar shapes that float before you like tiny bubbles.

This is no optical illusion! The shapes are tiny cloudings in the eyes, which throw shadows onto the retina. Since they are heavier than the liquid in the eye, they always fall farther down after each blink. If you lay your head on one side, the motes struggle towards the angle of the eye, showing that they follow the force of gravity.

216. Glowing eyes

Why do the eyes of cats, dogs, deer, and other nocturnal animals glow when light strikes them in the dark? The light is reflected by a special layer in the eyes of the animals. This layer consists of thousands of tiny crystals, which, incidentally, also cause the glimmer on fish scales. The crystals are situated behind the retina and act like a mirror, thus causing twice the amount of light to reach the optic nerve. That is also the reason why these animals can see well at night.

217. Moths that lose their way

Why do moths fly to the light? They are not attracted by it but misled. Flying at night, moths steer by the moon. They know they are flying in a straight line as long as the moon shines into their eyes from the same side. But when they pass a streetlight, for example, on the other side they get confused. They leave their straight flight path and approach in spirals in order to keep the light always on the same side.

218. Light on the beach

On summer nights, especially after rain, you may see the sea aglow with light. Sometimes it is a glimmer, sometimes the surface of the water shines white. The light is due to millions and millions of minute creatures, a species of flagellata (A), which you can catch in a glass jar. The light is produced by a substance that glows brightest when the supply of oxygen is most plentiful. That is why the tops of the waves and the breakers on the beach are especially bright, as are the marks of your footsteps or lines you draw with your finger in the wet sand, which bring air in close contact with the tiny creatures.

219. Stripes of light

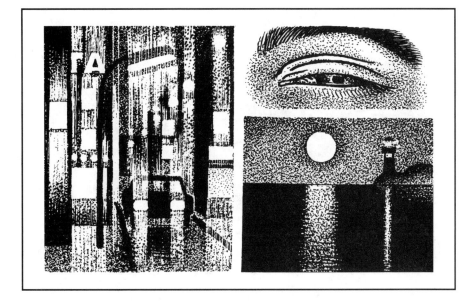

At night, when you look at the lights on the street with your eyes squeezed almost shut, the lights melt into vertical, bright stripes. When you tilt your head, the stripes of light slant—a sign that they are created by the eye.

The stripes are caused by reflections of the lamplight from the tear liquid along the edge of the upper and lower eyelids. The eyes see the bright lamps normally; the liquid, which stands directly in front of the pupil opening when the lids are half closed, reflects the lamplight in the form of long stripes on the retina.

220. Magic rabbit

Look at this picture at the normal reading distance. Then shut your left eye and stare at the magic rod with your right. If you now slowly alter the distance of the picture—abracadabra!—the rabbit suddenly disappears.

The retina of the eye consists of a large number of light-sensitive nerve endings, called rods and cones. There is one spot, however, where there are no rods or cones to detect light. It is located where the nerves join together to form the optic nerve. If the image of the rabbit thrown on the retina falls at this "blind spot" as you move the picture, you cannot see it.

221. The disappearing finger

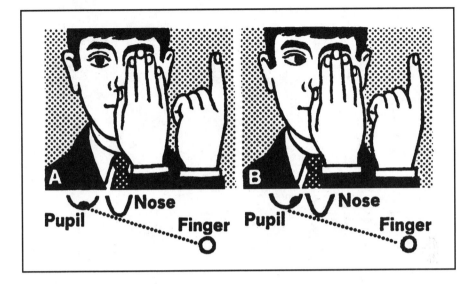

Cover your left eye with your right hand and look straight ahead with your right eye. Raise your left forefinger to your left ear and move it until the tip of the finger is just visible (A). If you now move your eye to look directly at the finger (B), strangely enough it disappears.

This interesting experiment has a geometrical explanation: when you are looking straight ahead (A), the light rays from the finger pass over the bridge of your nose into the pupil of the eye. But if the pupil is moved to the left (B), the light rays from the finger go past it.

222. Hole in the hand

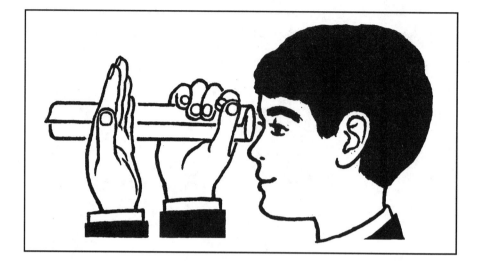

Roll a piece of writing paper into a tube and look through it with your right eye. Hold your left hand open on the left, next to the paper. To your surprise you will discover a hole, which apparently goes through the middle of the palm of your hand. Can you think what causes this illusion?

The right eye sees the inside of the tube and the left eye sees the open hand. As in normal vision, the impressions which are received by each eye are combined to give a composite image in the brain. It works particularly well because the image from inside the tube, which is transferred to the palm of the hand, is in perspective.

223. Moon rocket

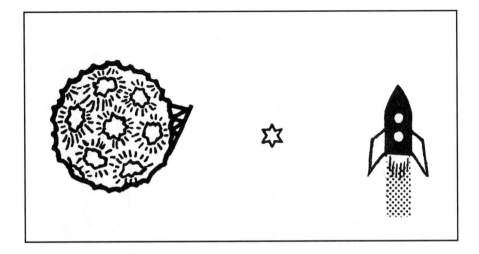

Hold the picture so that the tip of your nose touches the star, and turn it around slowly to the left. The rocket flies into the sky and lands again on the moon. Each eye receives its own image and both impressions are transmitted to the brain, which combines them. If you hold the star to the tip of your nose, your right eye only sees the rocket and the left eye only the moon. As usual, the halves of the image are combined in the brain. As you turn the picture on its edge, it does not shrink anymore because both eyes see the same image by squinting.

224. Ghostly ball

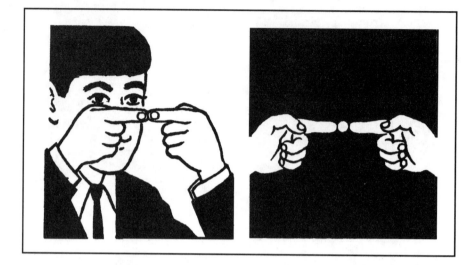

Hold your forefingers so that they are touching about a foot in front of the tip of your nose and look over the fingertips away to the opposite wall. On doing this you will see a curious ball, which is apparently fixed between the fingertips.

When you look over your fingers your eyes are focused sharply on the wall. But the fingers are then projected on the retina in such a way that the images are not combined in your brain. You see the tips of both fingers doubled. These finally combine to give the illusion of a round or oval image.

225. Two tips to your nose

Cross your index and middle fingers and rub them sideways over the tip of your nose. To your great surprise you will feel two noses.

When you cross them over, the position of the sides of the fingers is switched. The sides normally facing away from one another are now adjacent, and both touch the tip of the nose together. Each one reports separately, as usual, the contact with the nose to the brain. This is deceiving because the brain does not realize that the fingers have been crossed.

226. Reaction time

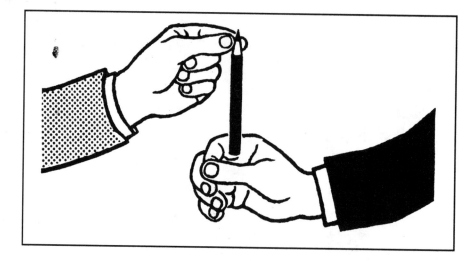

Hold a pencil above your friend's slightly open fist and ask him to catch the falling pencil by closing his hand. He cannot do it!

When the eyes see the pencil fall, they first send a signal to the brain, and from here the command "Grab" is sent to the hand. Time is naturally lost in doing this. If you try the experiment on yourself, it must succeed, because the commands to let fall and to grab are simultaneous. We call the time between recognition and response the reaction time. The time lost in a dangerous situation can mean death for a car driver.

227. Confused writing

Would you bet that you cannot write your name if you make circular movements with your leg at the same time? You will manage nothing more than an unreadable scribble.

It is probably possible to draw in the same direction as the circular leg movements. But as soon as you circle your leg in the other direction, the pencil movements cross over completely. So the leg movements are transferred to the writing. Each action needs so much concentration that both cannot be carried out at the same time.

228. Mistake in writing

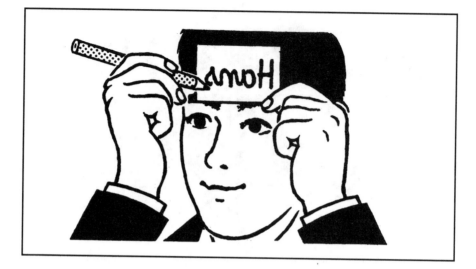

Hold a card in front of your forehead and try to write your name. You will be surprised at what appears. Your name is backwards, readable only in the mirror.

From pure habit you have started at the left and finished at the right, as you usually do when writing. This was a mistake, because if you had thought about it, the writing must be laterally reversed!

229. Big swindle

Put a bottle upright on the ground and walk around it three times. If you then try to walk straight towards something, you will not be able to do it.

The balancing organ in the inner ear has played a trick on you. A liquid starts to move in it when you turn your head. Small hairs are bent when this happens and report the process to the brain. This makes sure that you make suitable counter-movements. But if you turn quite quickly and stop suddenly, the liquid goes on moving. The brain reacts as if you were still turning, and you go in a curve.

230. Annoying circles

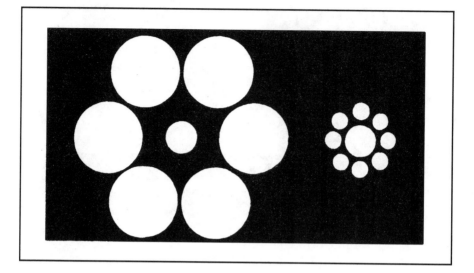

Look at these two figures. Which of the two circles in the middle of the figures is larger?

Both circles are the same size! In the subconscious mind we do not only compare the middle circles with one another, but also with those in the surrounding circles. In this way we get the impression that the middle circle in the right-hand figure is larger. We are subject to a similar optical illusion when we look at the moon. If it is close to the horizon, we compare it with houses and trees. It then appears larger than when it is high in the sky.

231. Magic spiral

Look at the picture closely. You will probably be certain that it is of a spiral. But a check with a pair of compasses shows that the picture is of concentric circles. The individual sections of the circles apparently move spirally to the middle of the picture because of the special type of background.

232. Measuring distance

Make a point on a piece of paper and place it in front of you on a table. Now try to hit the point with a pencil held in your hand. You will manage it easily. But if you close one eye, you will almost always miss your target.

The distance can only be estimated with difficulty with one eye, because one normally sees a composite image with both eyes and so can discern the depth of a space. Each eye stares at the point from a different angle (notice how the angle alters if you move nearer to the point). The brain can then determine the distance of the point fairly accurately from the size of the angle.

233. Crazy letters

In the picture, letters are embroidered on checkered material with yarn which is made of twisted black and white thread. Do you have any doubt that the letters are sloping? A ruler will show that the letters are straight. Because of the sloping bars in the background and the twisted threads, our eyes experience a confusing shift in the outlines of the letters.

234. Mirages

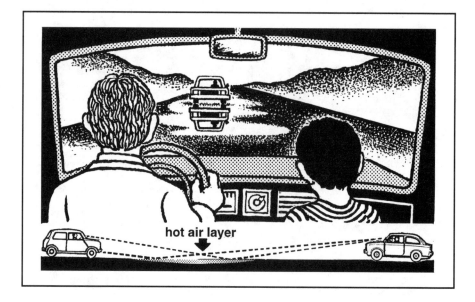

hot air layer

On hot days you may see reflections on asphalt roads that strikingly resemble water. How do they come about?

The dark asphalt absorbs the sunlight and heats up. Directly above the road lies a thin layer of hot, thinned air, which has a lower optical density than the cooler air above it. When light rays pass from one material into another that is optically denser and at a low angle, they are reflected and a natural phenomenon occurs which is known in desert regions as a mirage. In windy weather you see fewer of the reflections, because the hot air layer is blown away from the roadway.

235. Dangerous reflections

The more a house is surrounded by trees and bushes, the more often birds, will fly into the windows. Especially on the shady side, the reflections of the sunlit surroundings look lifelike enough to deceive the birds. The impact with the glass can kill birds, but may only stun them. A stunned bird should be put in a place safe from cats; it may come to after a while. Birds may be warned of the danger by a length of ribbon fastened to the outside of the window pane, where it flutters in the wind. Better still, cut the outline of a bird of prey, such as a hawk, from a piece of black paper and stick it on the window.

236. The turning phonograph

If you move the phonograph in the picture above in a circle, the record appears to turn.

The apparent movement of the record has many causes. When we move the picture, the constantly changing light incidence and angle of observation produce in the eye moving dark and bright zones, which apparently move across the record. The eye cannot follow so quickly, and sees normal rotation of the record.

237. Money fraud

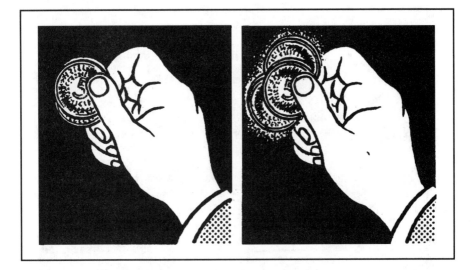

Hold two coins of the same value between thumb and forefinger, and rub them together quite quickly. If you look carefully, you will detect yet a third coin, which apparently moves backward and forward between the others! What causes this startling illusion?

Our eyes react too slowly to follow the rapid movement of the coins. Each time, the image of the coins remains for a little while on the retina, although they have already moved away. So we see both coins in movement and the after-image of a third coin.

238. Cinematographic effect

When you stand in front of a lattice fence, you can see only a little through its narrow gaps. But when you drive along the fence, it seems to be almost transparent. How can that be explained?

Our eyes are lazy; when we are driving along, the image that is seen through each fence gap persists in the retina for a little while, until the image passing through the next fence gap is impressed upon it. The individual image impressions melt together to form a coherent image, as in a film, where in one second 24 pictures pass by without any flicker. Since the eyes focus on light objects in the distance behind the fence, they perceive the dark laths, which quickly pass by in the foreground, only as a blurred surface.

239. Wheels in a film

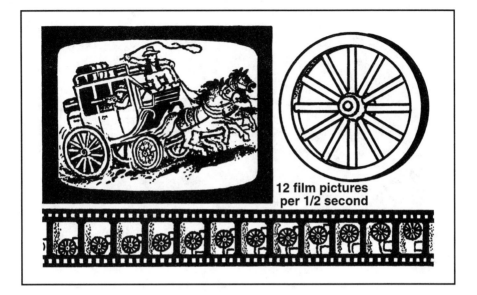

12 film pictures per 1/2 second

Sometimes, the wheels of a fast-moving stage coach or a car seem to be standing still or to be turning backwards. How is this illusion created?

During the filming, twelve pictures are exposed in half a second. When a wheel with twelve spokes revolves once every half a second, the spokes have the same position on all 12 pictures, so that when the film is shown in a movie theater it seems as if the wheel is standing still. One says the film pictures and the spokes run synchronously. When the wheels slow down a little, each spoke stays a little farther back on each picture and it looks as if the wheel is turning backwards.

240. Fast succession of pictures

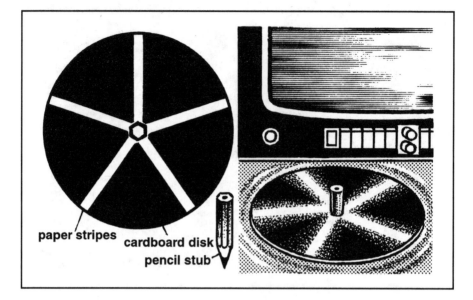

paper stripes   cardboard disk
pencil stub

You can easily observe the broadcasting frequency of your television with a piece of round cardboard covered with black paper and 5 white stripes. When you turn the top on a pencil in the dark room directly in front of the bright television screen, the stripes all of a sudden stand still, then they move backwards and double. Every second, 30 pictures are broadcast, and, after each picture, there is a short break when the screen goes dark. If the top makes exactly six turns per second, the succession of the pictures and the revolution of the stripes have the same rhythm. It then seems as if the stripes are standing still. When the top becomes slower, the stripes seem to turn backwards until they are illuminated each twelfth of a turn by a television picture, and then 10 stripes are seen standing still.

GIANT BOOK OF SCIENCE EXPERIMENTS

241. Living pictures

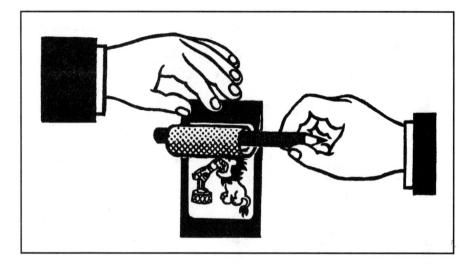

Copy and cut out the two filmstrips on page 258 and tape them together at the upper edge with number 1 on top. Roll up the top sheet and move it up and down with a pencil. You get the impression that the figure is moving.

The impressions of the pictures received by the eyes merge in the brain and produce the effect of movement. This "cinematographic effect" seems very primitive in this case because it is only produced by two pictures. In normal films 24 pictures roll by in a second, and in television as many as 30, so we see smooth and flicker-free movement.

242. Movies in a candy box

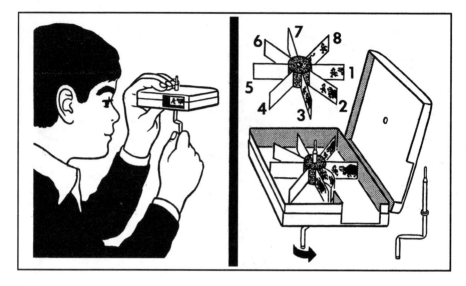

Copy and glue one of the filmstrips provided on page 258 onto drawing paper, separate the eight little pictures, and place them in the right order into a notched cork disk $5/8$ inch thick. Cut along the side wall of a folding candy (or similar) box for $1 1/4$ inches and stick two-thirds of it at right angles facing inwards. Color the inside of the box black and make holes in the middle of the base and lid. Bend the crank from a worn-out ballpoint pen shaft. Fix it into the box and place the cork, which you have previously bored, firmly on it. If you turn the crank around to the right, the figures move.

When you wind the crank, each little picture is seen by the eyes for a moment and is quickly replaced by another. Because the eyes are sluggish, each picture leaves an after-image when it has already moved away. The individual pictures merge into one another to give the appearance of movement when you wind the crank. This discovery was made in 1830, and today the most modern equipment in the cinema works on the same principle as your movie in a box.

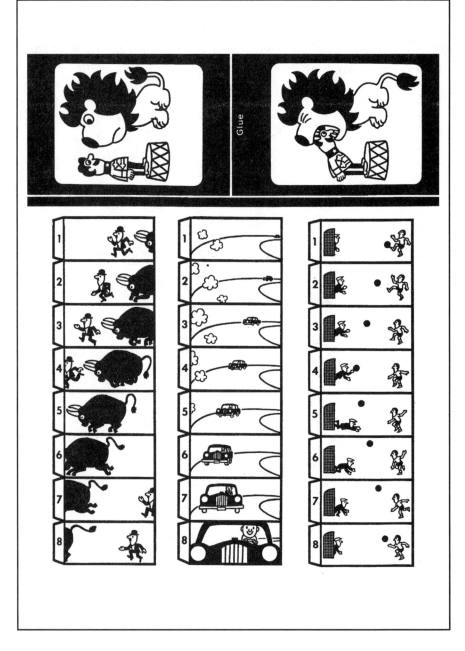

243. Effect of the sun

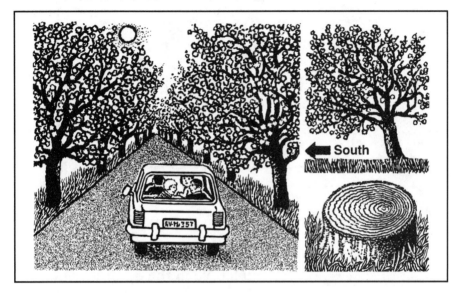

South

On a sunny spring day, a car is driving southward on a country road. A passenger admires the blooming apple trees along the edge of the road. "If you drove on this road in the opposite direction, the trees look even nicer," another passenger remarks. What does he mean?

Since the car is headed south, the passengers only see the northern side of the trees. But because the growth of free-standing trees is influenced by sunlight, they are better developed towards the south than towards the north. Accordingly, the annual rings in the trunk are wider on the south side because more tissue is needed for the increased supply of water and nutrients. This is clearly recognizable in a freshly cut tree stump.

244. Maze

Plant a sprouting potato in moist soil in a pot. Place it in the corner of a shoe box and cut a hole in the opposite side. Inside, tape two partitions (as shown in the illustration) so that a small gap is left. Close the box and place it by a window. After a couple of days the shoot has found its way through the dark maze to the light.

Plants have light-sensitive cells which guide their direction of growth. Even the small amount of light entering the box causes the shoot to bend. It looks quite white, because the green chlorophyll, necessary for healthy growth, cannot be formed in the dark.

245. Steered growth

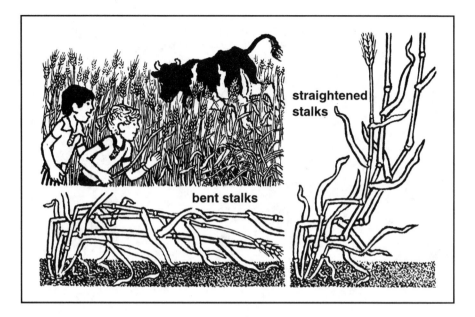

straightened stalks

bent stalks

"The damage is only temporary," a boy explained to his friend, whose cow ran into a rye field. "The bent stalks will straighten themselves." Is this true? Try it yourself and see what happens.

Wheat and other grains we eat belong to the family of grasses. Each of the stalks grows above the knots in the soft, light parts of the stem, which are surrounded and supported by tube-shaped leaf sheaths. If you bend a stalk, as long as it is still green, the growth at the knot above the bend is redirected by the sunlight. The leaf sheath begins to grow stronger on the side which is turned away from the sun and it bends the stalk back up until the head stands upright again in the air and sun after a few days.

246. The willow's secret

Why do willow trees often stand in a straight line along ditches and at the edges of fields? They probably grew from green willow sticks that someone put into the ground as a fence. Where the earth was moist, they made roots and grew into trees. When the broom-like shoots at the top of the posts had been cut back often enough, the trees formed thick trunks. If you place willow branches in a glass of water they will make roots and can then be planted.

247. The sun brings life

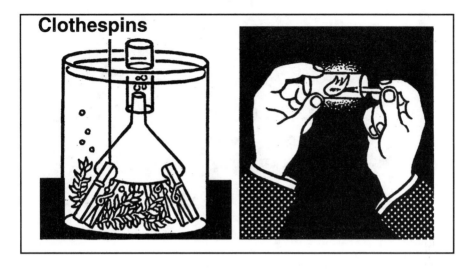

Clothespins

Fill a large glass jar with freshwater and place several shoots of waterweed in it. Place the jar in sunlight, and at once small gas bubbles will rise in the water. Invert a funnel over the plants and over it a water-filled glass tube. The gas which is given off by the plants slowly fills the tube.

Plants are powered by sunlight. With its help, in the presence of chlorophyll, they make their building material, starch, from water and carbon dioxide, and give off oxygen. Oxygen has actually collected in the glass tube. If you remove the tube and hold a glowing match in it, the match will burn brightly.

248. Automatic watering

Fill a bottle with water and place it upside down and half buried in soil in a flower box. An air bubble rises up in the bottle from time to time, showing that the plants are using the water. The water reservoir is enough for several days, depending on the number of plants and the weather.

Water only flows from the bottle until the soil around it is soaked. It starts to flow again only when the plants have drawn so much water from the soil that it becomes dry, and air can enter the bottle. Notice that plants can take water more easily from loose soil than from hard.

249. Secret path

Dissolve a teaspoonful of salt in a glass of water and cover it tightly with parchment paper. Place the glass upside down in a disk containing water strongly colored with vegetable dye. Although the parchment paper has no visible holes, the water in the glass and the dish is soon evenly colored.

The tiny molecules of water and dye pass through the invisible pores in the parchment paper. We call such an exchange of liquids through a permeable membrane osmosis. All living cells are surrounded by such a membrane, and absorb water and dissolved substances in this way.

250. Pressure from osmosis

At the edge of the road you will discover many different plants (such as grass, dandelions, and others weeds) which have arched up and burst through the asphalt surface. Where do these plants get such enormous strength?

The plant sprouts draw water from roots below the asphalt. The water is transported from cell to cell via osmosis, passing through the semi-porous cell walls. As the water fills the plant cells, their internal pressure increases until a force similar to that of a jackhammer is created.

Similarly, if you plaster a row of dry peas into a box and then water the hardened block of plaster, it will soon burst open due to the pressure in the cells of the peas.

251. A dandelion's strength

If you slit the stem of a dandelion into long strips and place it in a glass of water, the sections will roll up like watch springs. Why? The spongy cells inside the stem expand strongly as they absorb additional water. In an uncut stem the strong outer layer of cells prevents this. The high pressure of the inner cell layer pressing the outer layer (comparable to the air-filled tube in a bicycle tire) gives strength to the dandelion stalk.

252. Splitting cherries

In very rainy weather, ripe cherries will burst and split on the tree. The same thing happens when you place cherries in water. The tiny pores in the skin of the cherry allow water to enter but they do not allow the thick, sugary juice to get out. The water is drawn into the sugar molecules in the juice, which become more diluted. The extra water increases the pressure in the cells of the cherries until they cause the fruit to burst. The movement of liquids through the cell walls of a plant is called osmosis. It is the same process by which the plants, from the roots to the leaves, are able to absorb water.

253. Rising sap

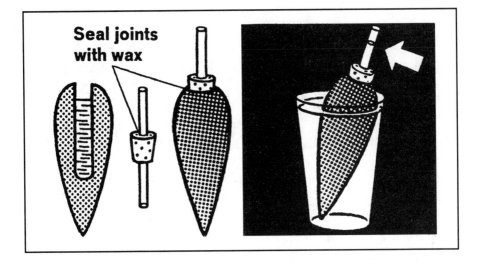

Seal joints with wax

Make a deep hole in a carrot and fill it with water in which you have dissolved plenty of sugar. Close the opening firmly with a bored cork, and push a plastic straw through the hole. Clean off any overflowing sugar solution, and seal the joints with melted candle wax. Put the carrot into a glass of water and watch; after some time the sugar solution rises into the straw.

The water molecules can enter the carrot through the cell walls, but the larger sugar particles cannot come out. The sugar solution becomes diluted and rises up the tube. This experiment on osmosis illustrates how plants absorb water from the soil and carry it upwards.

254. Ghostly noise

Fill a small goblet to overflowing with dried peas, pour in water up to the brim, and place the glass on a metal lid. The heap of peas becomes slowly higher and then a clatter of falling peas begins, which goes on for hours.

This is again an osmotic process. Water penetrates the pea cells through the skin and dissolves the nutrients in them. The pressure thus formed makes the peas swell. In the same way the water necessary for life penetrates the walls of all plant cells, stretching them. If the plant obtains insufficient water, its cells become flabby and it wilts.

255. Interrupted water pipe

bark | annual rings | sapwood | pith | cambrium

A man sets up a Christmas tree on Christmas eve. He must carve away the outer wood of the tree trunk, so that it fits into the stand. Even though the stand is then filled with water, the tree becomes dry and loses its needles shortly after the holidays. How can that be explained?

When the trunk was cut, the outermost annual ring—the sapwood—is removed along with the bark. In the sapwood long, stretched-out cells which act as tiny pipes transport the water (and nutritive substances) up to the branches and needles. The older annual rings towards the center of the trunk are composed of dead, wooden cells that do not transport the water. They keep the trunk firm to support the tree.

256. Rain in a jar

Place a green twig in a glass of water in sunlight. Pour a layer of vegetable oil on to the surface of the water and invert a large jar over the plant. After a short time, drops of water collect on the walls of the jar. Since the oil is impermeable, the water must come from the plant's leaves. In fact, the water, which the plant absorbs from the glass, is given off into the air through tiny pores in the epidermis, or skin, of the leaves. Air saturated with moisture and warmed by the sun deposits drops like fine rain on the cool glass.

257. Water from birch leaves

If you tie a plastic bag over the leaves on the branch of a birch tree, you will soon find moisture collecting at the bottom of the bag. The water comes from many tiny pores in the leaves. It condenses on the inside of the bag. On a hot day, the amount of water the leaves "transpire" will be quite large, although far more water is lost by the uncovered leaves surrounded by dry air than by those in the moisture-laden atmosphere of the bag. In summer a fully grown birch tree may give off as much as 100 gallons of the water it has taken up through its roots.

258. Nature in a bottle

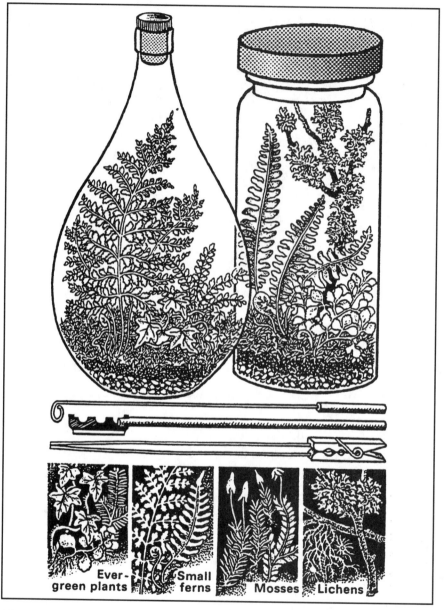

Evergreen plants · Small ferns · Mosses · Lichens

Plants can live for a long time without attention in an airtight, clear-glass bottle or jar. First, insert small pebbles and pieces of charcoal, using a funnel if necessary; next, add some soil mixed with peat and half a teaspoonful of bone meal. Add enough water to make the soil damp but not soggy. Choose only slow-growing, non-flowering plants, such as small ferns and mosses, and with the aid of some wire instruments place them on the damp soil.

Close the bottle firmly and keep it in a light place but away from direct sunlight. Open the bottle only occasionally to remove dry leaves and perhaps to add a little water.

In the "bottle garden" the food cycle functions almost exactly as in the open air: (1) The plants absorb water from the soil and evaporate most of it through their leaves. The moisture condenses in droplets on the glass walls and falls like rain to the ground. (2) The plants breathe in carbon dioxide and breathe out oxygen (the opposite of what we do). Some of the oxygen is absorbed again during the hours of darkness. The rest is absorbed by bacteria in the soil and by fungi which, in turn, produce carbon dioxide. (3) The plants need light to power the green chlorophyll molecules in the leaves, which manufacture starch (the plants' main food) from water and carbon dioxide. The starch is broken down into its original components as parts of the plant decay.

259. Zigzag growth

Lay pre-germinated seeds on a sheet of blotting paper between two panes of glass, pull rubber bands around the panes, and place them in a water container by a window. Turn the glass panes with the shoots onto a different edge every two days. The roots always grow downwards and the stem grows upwards.

Plants have characteristic tendencies. Their roots strive towards the middle of the earth and the shoots go in the opposite direction. On slopes the roots of trees do not grow at right angles to the surface into the ground, but in the direction of the middle of the earth.

GIANT BOOK OF SCIENCE EXPERIMENTS

260. Spruce or fir?

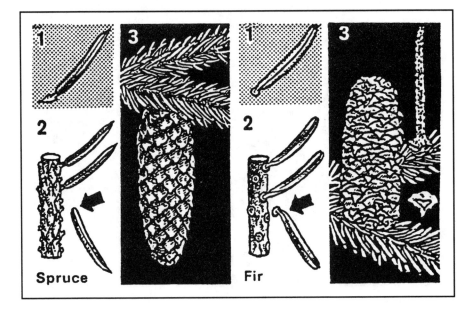

It is easy to distinguish between spruce and fir trees.
(1) Spruce needles are square and green all around. Fir needles are broader, with two light lengthwise stripes on the underside. (2) Bare spruce branches are rough like a rasp, because the little stems of the needles remain on the branch when the needles drop off. The disc-like stems of fir needles drop with the needle and leave the branches fairly smooth. (3) Spruce cones hang upside down on the branches and drop whole. Fir cones stand upright and lose their scales until only the bare axis is left before they drop.

261. Collecting leaves

Aspen

Aspen leap of a sucker

Poplar

Underside of the leaf

To make a collection of dried plants, start with the leaves of various trees. Take two pieces of hardboard about the size of a sheet of paper and sandwich between them a layer of newspaper about as thick as a finger, all sheets cut to the same size. Two strong elastic bands will hold the "sandwich" together. Collect leaves of a suitable size from all sorts of trees and place them to dry between the layers of newspaper for about two weeks. When thoroughly dry, the leaves can be stuck on a sheet of paper and labeled. Some leaves from secondary shoots arising from the trunk of a tree, called suckers, often appear different from the other leaves of the tree.

262. The life cycle of a tree

The rings visible on the stump of a cut tree show not only the age of the tree but also tell a tale of good and bad years. Broad rings indicate years of sunshine and rain; narrow rings indicate years of drought or cold. The light-colored "early wood" of a ring grew in spring and contained soft, sap-conducting cells. It merges with the "late wood" grown in summer and autumn before the tree stopped growing at the onset of winter. It has narrower pores and is darker and harder than the sapwood.

263. Crooked trees

Why is it that trees growing on a steep slope often have crooked trunks? The shape of the trunk shows that the ground has moved. Heavy rain causes the topmost layer of soil to slide downwards, and young trees gradually bend over. Because trees have a tendency to grow upright, however, the trunk bends immediately above the ground. The roots also turn to grow vertically into the ground. The tree is able to withstand the movement of the ground when it has reached a certain age.

264. Crippled trees

We are sometimes surprised by the peculiar shapes of trees which stand on their own, especially near the sea and in the mountains. They are crippled trees. Year after year, wind-borne sand and ice crystals wear away the young shoots to windward, or do not even allow the delicate buds to form. On cliffs by the sea you can tell the direction of the prevailing wind from the outline of a group of trees. If the prevailing wind blows from the sea, the trees bend towards the mainland and their windward sides have been shaved by gales.

265. Tree growth

growth in thickness / growth in length

Hammer a nail into the trunk of a tree to mark the height of a child. You might guess that the child would hardly grow, while the tree trunk would become considerably longer, so that the nail will be noticeably higher than the head of the child. But this is not the case.

The trunk of a tree with its wooden, dead cells can never stretch upwards. It does grow, though—as the branches and roots do—in thickness. In the growth layer (cambrium), a new ring is created every year. Only at the tips of the branches and roots does the tree grow upwards.

266. Leaf skeleton

Place a leaf on blotting paper and tap it carefully with a clothes brush, without pressing too hard or moving sideways. The leaf is perforated until only the skeleton remains, and you can see the fine network of ribs and veins.

The juicy cell tissue is driven out by the bristles and sucked up by the blotting paper. The ribs and veins consist of the firmer and cellulose-strengthened framework, which resists the brush.

267. Quaking aspen leaves

The aspen tree looks like a giant mobile when its leaves move in a gentle breeze. Depending on the direction of the wind, the round leaves on their long, flattened stalks move back and forth (1) or rotate, alternately showing their dark tops and light undersides and seeming to flicker (2). Or they move up and down. The leaves are slightly arched, and this causes the air pressure below them to decrease, so that the normal air pressure above depresses them a little (3).

268. Why leaves fall

Break a small branch on a tree. In summer the leaves on the broken branch dry up but do not drop off, even in the autumn when the tree loses all its other leaves. Why? In the fall leaves drop because a thin layer of cork forms at the upper end of the leaf stems between two layers of cells, blocking the flow of sap. The cork layer creates a weak point, allowing the dead leaves to break away. The leaf stems on the broken branch have not formed this layer of cork. The sap-conducting veins have dried up but are not blocked.

269. A hollow willow tree

The trunk of a willow tree may be so hollow as to consist only of a thin skin, but it still carries some leafy branches. The leaves will not wither, even if you peel off a ring of bark without cutting into the wood (arrow). These two observations prove that the sap of the tree is not conducted by the heartwood or by the bark. Only the outer sapwood of the youngest rings carries the "veins" through which the water can rise.

270. Tree chokers

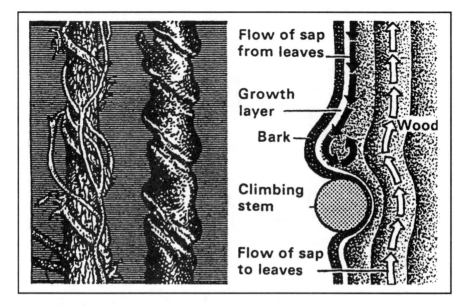

The trunk of a young tree overgrown with honeysuckle sometimes twists like a corkscrew. What is the reason for the swelling just above a climbing stem that has cut into the trunk? When the trunk increases in size, the woody stem of the honeysuckle will not give and chokes off the downward flow of sap immediately beneath the bark, thus cutting off the food supply from the leaves that the tree needs for growth. The sap is dammed above the blockage, and this causes the wood and the bark to grow faster there.

271. How the oak protects against decay

A split log of oak shows that the inner heartwood is darker than the outer sapwood. The dark coloring is due to a substance similar to that used by a tanner in preparing leather. This tannin, or tannic acid, protects the wood against bacteria, fungi, and insects. The darker the wood, the stronger the protection against decay. That is the reason why oak trees can reach an age of 1000 years. But a damaged tree, for instance one that has been struck by lightning, may rot because the outer protective layer of bark is gone. If you want to save the tree, cut out all the affected wood and fill the space with cement to prevent the tree from being broken by gales.

272. Holes in hazelnut shells

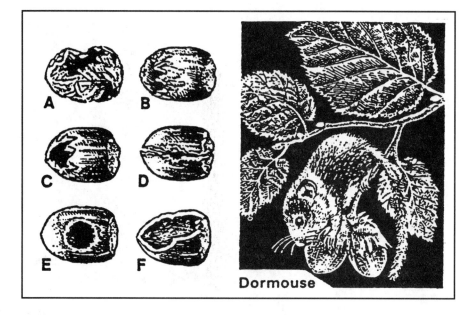

Dormouse

Each of the various animals which like to eat the sweet hazelnut kernels leaves its own mark on the shell. Young squirrels nibble the shell on all sides until it breaks (A). Older, experienced squirrels gnaw off the tip because there the shell is thinnest (B), or pry it open with their teeth (C). Sometimes squirrels cut a vertical groove around the nut and crack it (D). Circular holes with the traces of sharp little teeth around the edges are due to mice (E). The woodpecker jams a nut in a crack in the bark of a tree and chisels it open (F).

273. Telltale traces on spruce cones

Mountains of discarded spruce and fir cones, mostly pecked only at the tip, are heaped up below a woodpecker's "anvil" (A). This is a crevice in a tree trunk or tree stump into which the woodpecker wedges a freshly picked cone to extract the oily seeds with its bill. Frayed scales on a spruce cone are the work of a crossbill (B). It sits head down on a cone, hooks its crossed bill under the scales, and breaks them apart.

274. Gnawed cones

A shower of spruce cone scales comes rattling through the branches. When you look up, you discover a squirrel busily working away on a cone with its sharp teeth. It rips off the scales one after the other to get at the seeds underneath. Finally it drops the remaining core, or axis, with a few scales left at the top (A). Wood mice, on the other hand, strip the cone to the last scale (B). They climb to the very top of the tree where the cones are.

275. Teeth marks on trees

When animals nibble the bark of trees and leave teeth marks on trunks and branches, they are not always very hungry. The red deer stag "frays" long strips of bark, especially from spruce, ash, and beech (A), while fallow deer are content to nibble (B). The hare, too, takes strips of tender green bark (C), and the rabbit bites right into the sap-wood; the traces of its upper teeth are more marked than those of its lower ones (D). The squirrel peels off the bark in a spiral (E), and the dormouse leaves small dents (F).

276. A pollination experiment

On mild days in early spring, some of the buds on the
hazel bush have tiny dark red threads, or styles (arrow).
These are its female flowers, which the wind pollinates with
the yellow pollen from the male flowers, the tassel-shaped
catkins. Spread a little of the yellow powder (if possible,
from a catkin of another hazel bush) on the red styles and
mark that branch with a piece of string or ribbon. Note
how the flower changes after pollination, and watch the
hazelnut slowly developing from the thicker part at
the base.

277. Sleeping flowers

The delicate flower of the wood anemone is carried erect and opens only in sunny, mild weather. It reacts more quickly than other flowers to darkness and light. When a cloud obscures the sun, or if you cover the plant with paper or a tin for a few minutes, the flower closes and the stem bends down. When the sun shines, increased water pressure in the stem and at the base of the flower makes the stem straighten up and opens the flower. The water pressure is increased by chemical reactions in the plant, which occur only when there is enough light and warmth.

278. The irritable wood sorrel

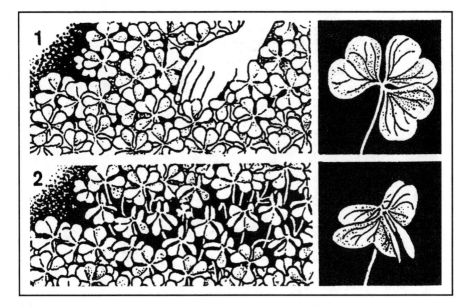

In the spring, wood sorrel often forms a bright green carpet in deciduous woods. It acts very strangely. Brush your hand over the leaves, and they slowly begin to fold. In a way which is not yet fully understood, the pressure in the cells at the base of the leaves decreases and the little leaves droop as though moving on hinges. Darkness and excessive sunlight have the same effect on the leaves of the wood sorrel.

279. Flowering while you watch

On warm summer evenings you may watch the opening of the evening primrose, which is found along the edges of fields and woods. It attracts moths at night. Keep your eyes on a bud that has its sepals (the outer ring of green petals around a flower bud) already beginning to open (1). Suddenly they open wider, and, as you watch, the bright, sulfur-yellow petals of the flower unfold and the sepals jerk back and downwards (2). All this happens within three minutes. The next day, that flower has withered already, and a new one opens up.

280. Two-colored flower

Dilute red and blue fountain pen inks with water and fill two glass tubes each with one color. Split the stem of a flower with white petals, such as a dahlia, rose, or carnation, and place one end in each tube. The fine veins of the plant soon become colored, and after several hours the flower is half red and half blue.

The colored liquid rises through the hair-fine channels by which the plant's water and food are transported. The dye is stored in the petals while most of the water is given off to the air.

281. Insects and the color of flowers

Bees, hovering flies, and butterflies like to rest on wash hung out on a line in the sun. When you watch them you will find that they prefer white and pale colors. They show the same preference in flowers. Bright red will attract only butterflies; other insects are unable to see red. This is particularly noticeable in the woodland. Strong or dark colors, such as red, purple, and blue, look even darker in the shade, and this explains why the flowers on the trees there are almost exclusively white and pale pink. Other colors were overlooked by the insects which might have pollinated them, so they did not form seeds and died out.

282. How many slices?

Before peeling an orange, you can tell someone how many slices it contains. This trick is accomplished by removing from an orange the small green or brown part of the plant that once formed at the bottom of the orange blossom. Within the hole left behind is a circle of small dots. Each dot indicates one orange slice; a very small one hints at a thin slice.

The dots are the remains of the seed vessels in the orange blossom, which are arranged in a circle. After being pollinated, the seed vessels develop into the orange slices, which form the fruit.

283. Green patches in the pasture

In summer, cows often bite off the grass right down to the roots but leave untouched lush patches of juicy green grass. "Even when they are hungry, cows will not eat the grass that grows on their own manure," says the farmer. Where the cows dropped their dung in the preceding year, the tender grasses have been killed off by the strong manure, and coarse grass has grown in their place, which the cows do not like.

284. The common butterwort

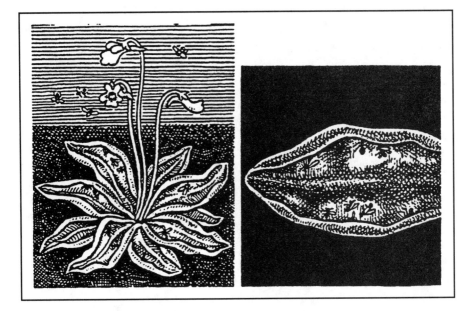

The common butterwort is able to grow in boggy meadows and rocky crevices which offer hardly enough food to other plants, because it adds meat to the food on which plants ordinarily live. A great number of glands in its leaves secrete a mucus which traps flies, mosquitoes, and other little insects. When the leaf has caught a fly, it oozes a liquid similar to the digestive juice of mammals, which dissolves the insect, leaving only the hard outer skin. The leaf then absorbs the nourishing "soup" to the benefit of the plant.

285. Feeding the sundew

The sundew grows in boggy places, where the soils have few of the nutrients the plant needs. To make up for this, the plant has sweet-smelling, glittering droplets on the red hairs edging its leaves, which attract small, nutrient-rich insects. As soon as an insect lands on what it mistakes to be honey, it is trapped. The sticky hairs and the leaves envelop it and gradually dissolve it with a digestive juice. You may feed the sundew with cheese, meat, and egg. Place some tiny crumbs on a leaf, and in one or two days they will have been "eaten." The sundew, however, does not react to bread crumbs.

286. A plant hunting underwater

If you see golden-yellow flowers resembling snapdragons floating on the surface of ponds or ditches, they may be the tops of a strange plant called bladderwort. It catches little creatures in the water. The plant has no roots and you may carry it home in a glass jar filled with pond water. On the underwater branches of the bladderwort you will see a number of tiny sacs. These are the traps. Each one has a small trap door that opens to admit any little water creatures that may bump into it. The bladderwort derives its food from them.

287. The age of lichens

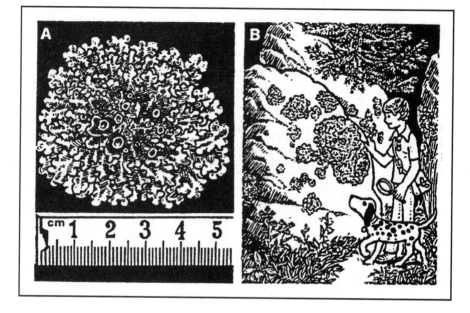

Make a note of the size of a particular patch of lichens and watch it spread in the course of the year. The yellow and greenish, usually circular lichens that grow on the bark of trees and on planks and stones may increase their size by $^2/_5$ inch, or 1 centimeter, annually (A). You may thus determine the age of lichens with the help of a tape measure. The lichen which forms green covers on rocks in the mountains (B) grows only by $^1/_{25}$ inch, or 1 millimeter, per year.

288. The decay cycle

The ground cover of leaves in a beech forest has three distinct layers. (1) The topmost layer consists of recently fallen leaves, some of which show traces of nibbling insects. (2) Below (from one year earlier) is a layer of partially rotted leaves which serves as winter quarters for insect larvae and pupae. (3) Still lower down (from two years earlier) is a crumbly layer of rotting leaves. This layer of leaf mold covers the roots of the trees that take food from it, helping the trees grow.

289. How ducks defy the cold

Top feathers
Tail gland

It really is amazing that ducks and other waterfowl are able to spend hours in ice-cold water. How do they protect themselves against the near-freezing water? They press their bills against a gland above the tail which makes a greasy substance like wax. They spread this over their feathers to make them water-repellent. You can prove this by sprinkling a few drops of water on a feather; they roll off instead of soaking in. Warm air is trapped by the soft down beneath the water-repellent top feathers. Thus the bird's body is insulated against the cold and floats on the water as though it were wearing a life jacket.

290. Birds in the wind

When icy winds blow over the seashore, you will see thousands of seagulls gathered on the ice and on the beach. As if by command, they all sit facing the wind. Why do they do this? The streamlined shape of the bird's body offers the least resistance to the oncoming wind (A). A strong wind from the back or side could upset the gull, ruffle its top feathers, and allow the warm air underneath to escape.

291. A bird box at the window

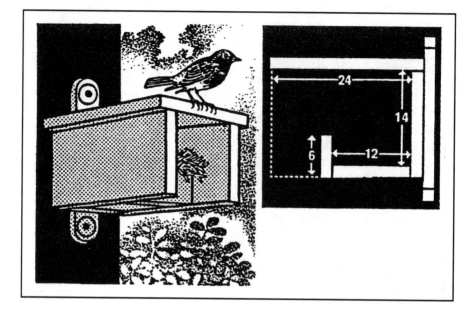

Redstarts, wagtails, and some other birds prefer partially open nesting boxes. Build the box from boards that are 1 inch thick. The nesting space should be about 5 inches square and protected against rain by an overhanging roof. At the back of the box fasten a piece of board by which you can attach the box to a wall or a tree. The box should hang about 9 feet above the ground, with the entrance facing south-east, near a window, so that you may watch the birds bringing up their young.

292. How to scare magpies

A magpie's eye is attracted instinctively by shining objects. From the top of a tree a magpie looks around for any round, light-colored things, such as the eggs of a songbird, which it steals out of the nest. Take one of the bright balls used for Christmas decorations, close it at the top with a waterproof adhesive, and fasten it to a post or a tree in the garden where it may be seen from all sides. The reflection of the sun on the ball will follow the suspicious magpie into every corner of the garden and irritate it so much that it will fly away.

293. The lizard's escape

The new-grown tail

When you chase a lizard, it will appear to automatically change direction as you stretch out your hand to catch it, just as we blink our eyelids when anything comes near our eyes. Your hand cannot keep pace with the lizard's zigzag path. Our seeing, thinking, and gripping takes a fifth of a second longer than the reflex movements of the lizard. Owing to its amazing nimbleness, the lizard's enemies often get hold only of its tail, which snaps off and, by continuing to wriggle, distracts the pursuer's attention. The wound heals quickly, and within a few weeks the lizard grows a new tail.

294. A frog's love song

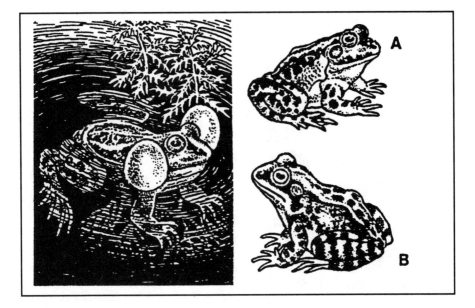

Frogs croak not through their mouths but by expelling air through their nostrils. The sacs that some frogs have in both cheeks and some at the throat amplify the sound, rather like the loud noise you hear when you rub an inflated balloon with one finger. The water frog (A) begins its croaking in the quiet hours of the evening with "moarks-moarks," followed by a continuous "breck-breck-breck." The grass frog (B) utters a gentle growl.

295. From egg to frog

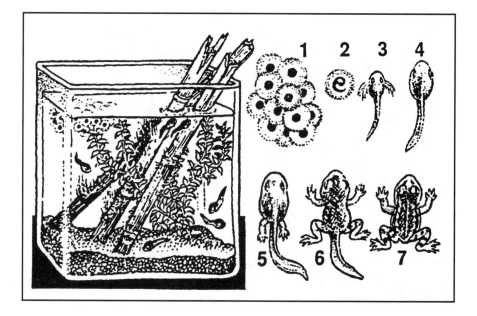

Put a little lump of frog eggs, dark side uppermost, in a
well-planted aquarium with clear pond water. Reed stalks
overgrown with algae or some stones should break the
surface. The little tadpoles (3) breathe through gills and
need some shade. With their horny jaws they nibble the
jelly-like remains of the eggs and the algae; later they will
eat bread crumbs, daphnia, and mosquito larvae. The fully
developed frogs breath through lungs. They climb out of
the water and should be removed from the aquarium.

GIANT BOOK OF SCIENCE EXPERIMENTS

296. Green frogs in the garden pond

A few small water and grass frogs may be kept in a large terrarium containing a spacious water basin and a variety of plants. As water frogs need water, they will also stay in a garden pond. They sun themselves among the plants and leap for insects, spiders, worms, and slugs, which they catch with their long, sticky tongues. How to catch live flies for your frogs? Make a funnel of wire gauze. Cut a few holes in the lower edge and place the funnel over a piece of raw meat. Flies will be attracted by the decaying meat and collect in a plastic bag tied over the top of the funnel.

297. A newt's growth

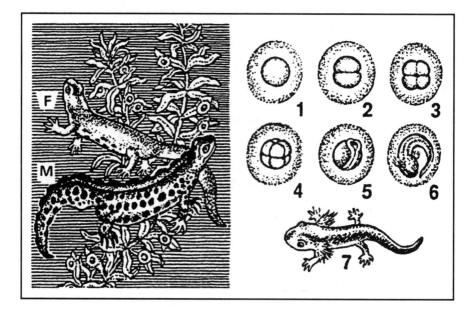

In the spring, you may find newts in a pond. The colorful male (M) attracts a female (F) and she attaches her eggs on water plants. You can keep newts for a few days in a covered aquarium and watch the eggs develop. Within three hours, a single egg cell divides into two, soon into four, eight, sixteen, and so forth. The larva emerges from the egg and breathes through tuft-like gills on each side of its head. Later these disappear, and the adult newt breathes through lungs. Feed the newts on daphnias and mosquito larvae.

298. Underwater magnifying glass

Construct an observation tube to get a clear and magnified view of animals and plants underwater without the usual reflections and distortions caused by the movement of the water. Cut off the bottom of a large can with a can opener and close the opening with a piece of transparent plastic, so that no water can get inside. When you insert the tube vertically in the water, the pressure of the water will slightly depress the plastic. Thus the light rays are deflected by the lens-like surface, as in a magnifying glass, and give the observer a magnificent view of the world underwater.

299. An artificial pond

A small pond in the garden attracts birds, which like to drink and bathe; it also gives you a chance to watch a variety of water creatures. Dig the pond in partial shade behind a bush. Line the edge with tiles to make it firm. Then line the hollow with a plastic sheet (you can buy these in garden centers and shops) and tuck the edges behind the tiles.

Get some bog plants and a bucketful of mud and set the plants along the edges. Do not plant reeds, as the shoots will pierce the plastic sheet. On the bottom of the pond spread washed sand from the beach, and on the edge lay some tiles and big stones. Leave an overflow in one place, where the water can run away and trickle into the ground; here you could plant irises. The water comes from a tap through a plastic hose covered with earth and runs into the pond through a little grotto of stones. Diving beetles and whirligigs soon will appear. In the winter, you may leave the water in the pond, but frogs and fish should be taken to deeper waters.

300. Making an aquarium

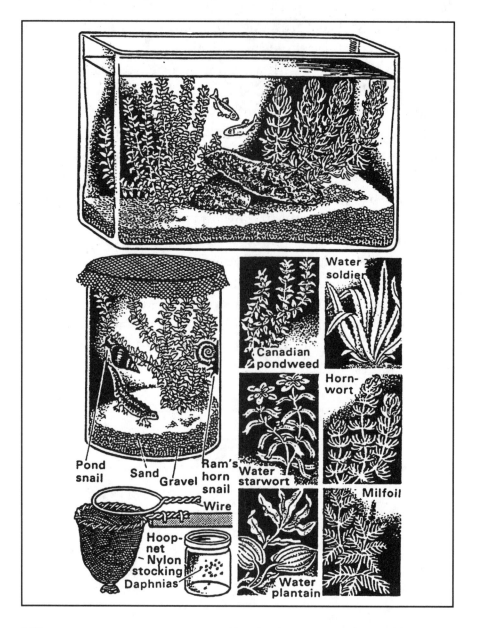

Water soldier

Canadian pondweed

Horn-wort

Pond snail   Sand   Gravel   Ram's horn snail   Water starwort

Milfoil

Wire

Hoop-net

Nylon stocking

Daphnias

Water plantain

It is quite easy to make a cold-water aquarium. You need a clean glass basin with a 1-inch layer of gravel on the bottom. Put groups of water plants and a mound of well-washed sea sand on the gravel. Carefully (over a piece of paper) fill the basin with tap water.

In an aquarium, as in a lake, the "biological balance" must be correct. There must be enough plants to provide all the oxygen the animals need to breathe. One fish the length of a finger requires about 4 pints of water. Decaying substances deprive the water of oxygen, causing the fish to die, as they do in a polluted lake. Dead plants should be removed and some of the water replaced by fresh water every two weeks.

For food, catch daphnias, mosquito larvae, and other small creatures in a pond. Feed the inmates of the aquarium twice a day, morning and evening, but give them no more than they can eat at one meal. The aquarium should be placed near an east or west window. The ram's-horn snail will keep algae down, but the pond snail eats the leafy plants.

301. Sticklebacks in an aquarium

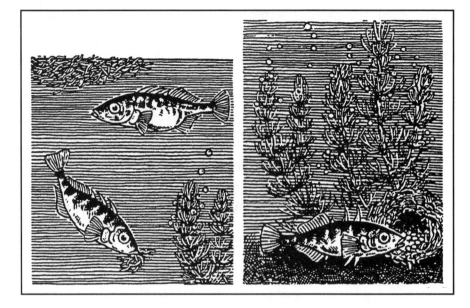

In spring, the male three-spined stickleback has glittering eyes, a vivid red belly, and a bluish-green back. If you keep one male and two or three females in your aquarium, you may watch their social behavior. Grass roots scattered on the water are taken by the male to the bottom, where he builds a nest. When the female has laid her eggs, he guards the nest, and fans water through it to aerate the eggs. These hatch in about 10 days, and any young fish that strays from the nest is quickly brought back in the mouth of its father.

302. Defending the territory

The male stickleback furiously attacks anything that may approach the nest. Even snails have to retreat and caddis fly larvae are removed by force. As the male drives away even his own females, these must be taken out of the aquarium. A dummy stickleback, made from blue and red modeling clay and lowered into the water at the tip of a knitting needle, is attacked just as ferociously as a willow leaf at the end of a twig. If you move the twig through the water, the stickleback will sometimes grip it so firmly with his teeth that you can pull him out.

303. A treat for butterflies

The damaged bark of a birch tree oozes sap, a sweet-smelling juice that attracts beetles and swarms of butterflies. They all love to drink from the fermenting birch juice—which makes them quite drunk. You can prepare a similar mixture from apple sauce, stale beer, and a little syrup and rum. Smear it on a tree trunk and fasten a piece of white paper below as a beacon for the insects at night.

304. A butterfly on your finger

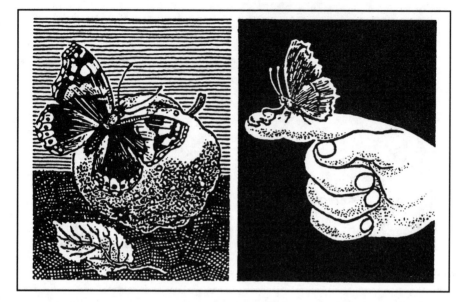

Butterflies often visit rotting fruit to drink the fermenting juice. This sometimes makes them so dazed that you can lift them by gently pinching the underside of their wings gently between two fingers. Never touch the wing tops or let them flutter in your hand. The wings are covered with dust-like color scales and are very easily damaged. Some butterflies will alight on your finger if you put a little jam on it, and will sit quite still. The insect will unroll its proboscis (sucking tube), which may be over half an inch long, dip it in the jam, and drink.

305. Houseflies and disease

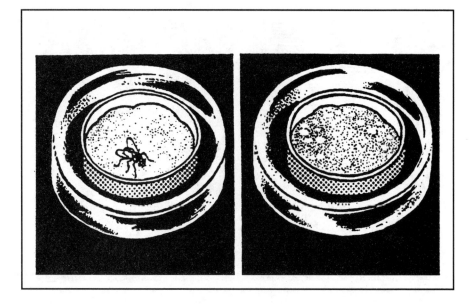

You can prove that flies can transmit diseases to our food.
Dissolve a small soup cube in hot water and thicken it with
a little corn flour. Put the mixture in a clean tin lid, cover it
with a piece of glass, and leave it to cool. Catch a fly and
put it under the glass. Let it walk over the cold mixture for
a minute or two and then release it. Cover the lid again.
After a day or two you will find small milky spots developing
where the fly has walked. These are colonies of bacteria
which the fly has introduced and which will multiply
rapidly.

306. An anthill

If you have lost your way in a pine forest, anthills will tell
you which way you are facing. They are usually found on
the south side of trees. The ants build their hills where they
are sheltered from rain and get the maximum amount of
sunshine. The hill often consists of dry pine needles, small
branches, moss, and earth. The hill conserves the heat of
the sun, which strikes it and allows a good airing of the
inside galleries. The pine needles on the outside are
arranged like tiles on a roof, so that rainwater cannot
penetrate. The nest itself is about as deep in the ground
as the hill is high.

307. More anthills, fewer pests

Queen Worker Male

Ants keep trees free of pests. Foresters know this and like to see as many anthills as possible. They even increase the number of nests, but only of the small red ants which, in contrast to the larger variety, have many egg-laying queens to each tribe. In spring when the ants come into the open to sun themselves, the forester scoops up a tinful of ants and puts them down beside a rotting tree stump on a pad of dry pine branches. He covers them with a layer of pine needles mixed with sugar and with more branches. The ants dig into the soil through the stump and establish a new nest.

308. Ant trails

Ant trails, which lead from the nest in all directions far into the forest, are marked with the specific scent of that particular tribe. Individual ants can produce chemical markers that create a trail others can follow. The ants' sense of smell is located at the tips of their antennae. They bend them to the ground to make sure that they are on their own chemical trail. When you brush your hand across the trail and disturb the chemical scent, the ants become confused. From far away, even from the tops of trees, the ants carry pine needles, bits of wood, caterpillars, butterflies, seeds, and beetles to their nest. A single load may weigh 100 times as much as the ant itself.

309. Ant farm

Nylon stocking

Paper

Elastic band

Moss

Glass tube

If you want to watch ants closely, dig out a colony of the small ants often found under stones in the garden. Carefully place the whole nest with the surrounding earth in a glass jar, cover it with a piece of nylon stocking, and shade the bottom part with paper. From time to time you may take the paper off to look at the underground galleries and watch the nursing of the larvae and pupae. Feed the ants pieces of fruit, sugar, and dead insects. Put a branch with some aphids in the jar and see how the ants touch them with their antennae and drink the sweet fluid they produce.

GIANT BOOK OF SCIENCE EXPERIMENTS

310. Noises in a nut

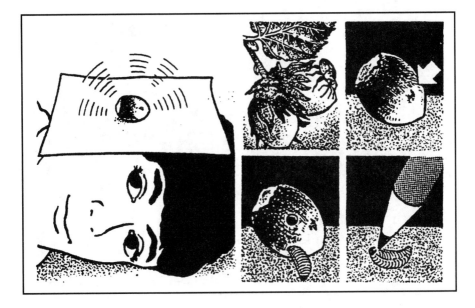

A small scar on a green hazelnut (arrow) is a good indicator that a hazelnut weevil (a snout beetle) is present. In spring, the female weevil drills a hole in the soft shell by driving her long tongue into it and walking around and around, and lays an egg in the kernel. If you stick one of these scarred nuts on a smooth piece of paper and hold it up to your ear, you will hear the greatly magnified noise of the larva feeding on the insides of the nut. It eats the whole kernel, and in September drills a small exit hole through the shell. It shows its sharp mandibles when you tease it with the point of a pencil.

311. The aphid's enemy

The seven-spotted ladybug is a most useful beetle in the garden, as you will find when you put it on a plant attacked by aphids. It quickly devours a few of the pests and is not at all scared by the ants which protect them. (Ants milk aphids like cows because of the sweet substance they produce.) The ladybug larvae, grayish-purple with black and yellow spots (A), also live on aphids. Put a branch with larvae in a glass jar (B) and watch the emergence of a ladybug. These little beetles also like to nibble a lump of sugar moistened with water.

312. A house of foam

On some wild flowers, you often find little lumps of white
foam. When you examine the interior of this foam, which
resembles spittle, you discover a little larva, well protected
from hungry insects, birds, and the heat of the sun (A).
The larva drills a hole in the stem of the plant, adds a juice
to the sap, and mixes them to form a soapy liquid which
makes bubbles when the larva blows air into it. Put the
larva on another similar flower and it will build a fresh
house of foam. The larva develops into an insect
something like a grasshopper: the spittle bug (B).

313. The firefly's signal

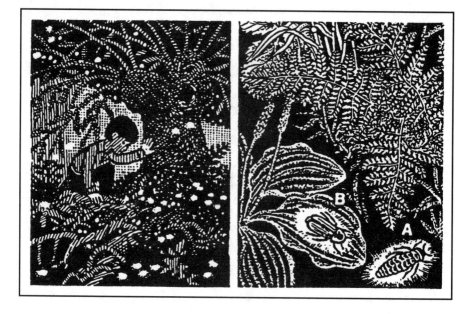

On warm June nights there are countless small lights flickering in the air and in the grass of damp fields and woods. With greenish light signals, the wingless female of the firefly (A) attracts the male (B), which also glows. When you catch one you will find that its light goes out immediately and comes on again only some time after you have released it. The cold light is produced in the stomach of the firefly by the oxidation of a substance called luciferin. When the male approaches the female, their lights get brighter, due to faster breathing and an increased supply of oxygen.

314. Flying spiders

On September mornings, you may see shimmering gossamer threads floating in the breeze. Snag one, and with a magnifying glass you can detect a tiny creature at the end of the thread. It is a young spider, looking for winter quarters. Spiders climb plants, fences, or walls and produce the threads from the spinner glands at the tips of their abdomen. The wind catches the silky threads along with the little spiders and carries them away, sometimes only to the nearest plant, sometimes over long distances.

315. A natural lawn fertilizer

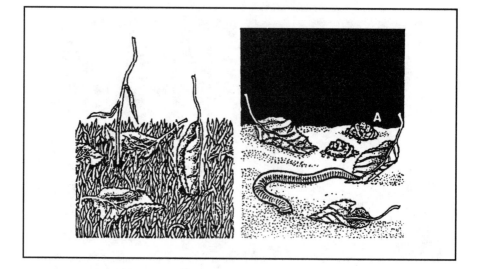

On lawns you will sometimes see rolled-up leaves and dry grasses standing upright with their tips buried in the ground. What makes them do that? At night earthworms come out of their underground passages and search for leaves and grass left lying on the lawn after it has been mowed. They take them into their mouths and pull them into the ground. When the leaves have decayed, the earthworms eat the remains, together with any earth clinging to them, and within their bodies the worms process the leaf matter and excrete it. The result are wormcasts (A), which help make a rich, fertile soil.

316. The earthworm's instinct

When you stick a piece of wood into the ground at an angle and tap it gently with your fingers, earthworms will come out all around you. Do they leave the ground to escape being eaten by their enemy, the mole? Or does the vibration of the soil indicate to them that rain is falling outside? Both of these reasons may be correct. Rain forces the worms to emerge because it fills their underground passages with water, so that they cannot breathe.

317. A natural protection

A blackbird poking its bill into the ground to catch an earthworm will try to pull it out. But earthworms have ways of escaping. Examine a worm closely with a magnifying glass and you will see stiff bristles on its belly. These serve to anchor the worm in the soil. If the worm snaps, the remaining part disappears underground. If it was the head piece it will live and form another tail. In spring you may see earthworms with a thick, orange-colored belt around the middle. Birds will not eat these, because apart from holding the earthworm's eggs, the belt contains a dangerous poison.

318. A creature with tentacles

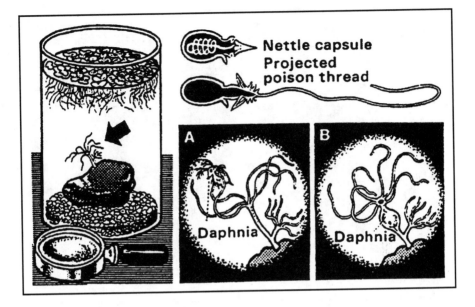

Nettle capsule
Projected
poison thread

A Daphnia

B Daphnia

If you keep water and duckweed from a pond in your aquarium, you may find a tubular creature less than an inch long, with thread-like tentacles that it moves in and out like the fingers of a glove. It is the brown hydra, a relative of the sea anemones and jellyfish of the ocean.

Unfortunately, with the naked eye you cannot see the nettle capsules containing minute poisonous harpoons which the hydra shoots at its prey. They can kill a daphnia (A). The tentacles wrap themselves around it and stuff it into the opening which serves as their mouth (B). You can see the hydra multiply by budding.

INVERTEBRATES

319. The life of a snail

Young Roman snails

Roman, or edible, snails seek shade and moisture between stones and under plants. In May or June, the snail digs a hole in the ground, about 4 inches deep, lays 60 white eggs the size of a pea, and covers the hole. Four weeks later, young snails with transparent shells will emerge. You may watch the growth of the shell by putting a dot of waterproof ink on the lower front rim. Soon you will find that new rings have formed. The shell consists of a lime-like substance secreted by the snail.

320. A snail's pace

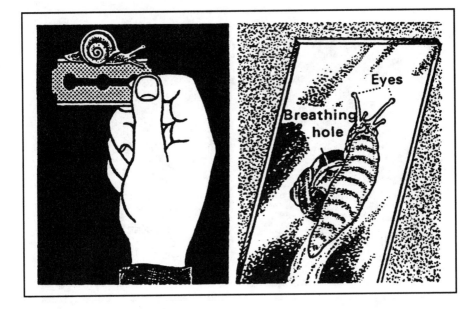

The numerous glands in a snail's foot secrete a slime on which it literally floats along. This efficient "greasing" allows a snail to crawl even on the edge of a sharp razor blade without damage to itself. Place a snail on a sheet of glass and watch the underside of its foot. You will see a number of shadowy stripes moving from back to front at an even pace. They are caused by wave-like contractions of the muscles, which keep pulling a piece of foot from the back and pushing it to the front. The speed of a snail is about 5 inches a minute.

321. Sweet snail food

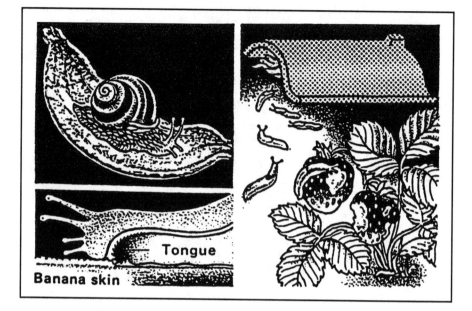

Tongue

Banana skin

Place a snail on the inside of a banana skin and it immediately begins to feed. Its tongue is like a rasp covered with thousands of tiny, backward-pointing teeth with which it scratches off the white layer of the banana skin. Snails like tender leaves and especially sweet fruit. At night, the little yellowish-gray slugs eat vegetables and fruit in the garden, making pests of themselves. They hide by day. You can attract them with potato peelings placed under a roof tile.

322. A pattern of spores

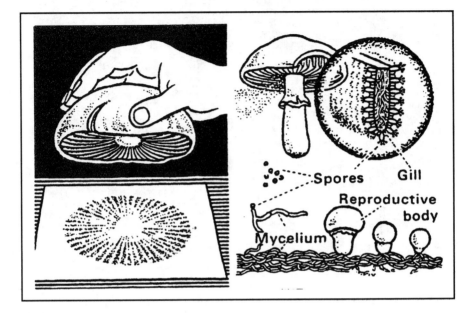

Place the cap of a gill- or tube-bearing mushroom on a
piece of paper in a warm dry room and it will produce a
pattern formed by fine dust. The dust consists of millions of
microscopic, single-cell spores which drop out of the tubes
or gills. The mushroom cap is the part of the fungus that
enables it to reproduce; the bulk of the fungus remains
underground and consists of a white network of threads,
called the mycelium. It develops from a single spore
carried by the wind to suitable ground. Years later it will
form new fungi.

323. Fairy rings

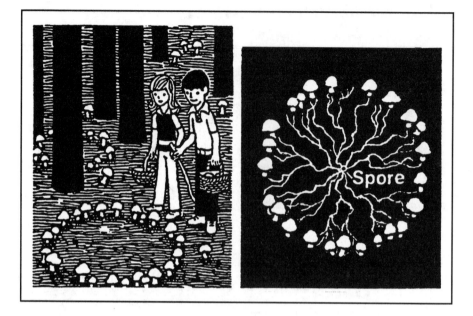

You may find mushrooms growing in "fairy rings" in the woods and fields. Once people believed that they marked the meeting and dancing grounds of witches. But the ring shape is due to the growth of the mycelium (network of threads from the fungi) beneath the ground. When a spore begins to grow, the mycelium forms a cobweb of threads spreading in all directions. Some years later, the oldest part of the mycelium in the center dies off, but the younger parts on the outside continue to spread from year to year to produce a ring of new fungi.

GIANT BOOK OF SCIENCE EXPERIMENTS

324. A special relationship

If you gently remove the earth under a pine tree where bolete mushrooms grow, you will find the small roots of the tree closely covered by the felt-like white mycelium of the fungus. The tree and fungus live together to their mutual advantage. This kind of permanent association between two plants is called symbiosis. The mycelium threads penetrate the outer layers of the roots and provide them with water and minerals. The tree, in turn, shares with the fungus some of the foodstuffs produced by its leaves.

325. The mole

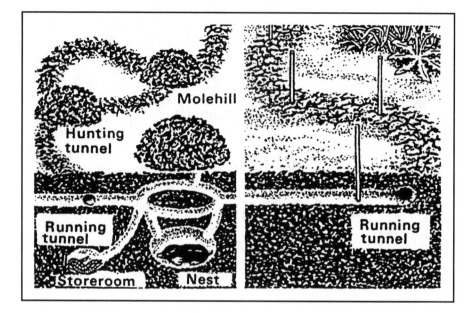

Within a few hours moles can dig an intricate network
of underground running and hunting tunnels radiating
outward from their nests. At about 8 o'clock in the morning
and again at 4 o'clock in the afternoon, moles run through
the hunting tunnels searching for earthworms, insect larvae,
and other creatures, and throw up fresh molehills. If you
carefully drill holes at intervals into one of the tunnels and
put a straw in each, you can see by the jerking of the straw
how quickly the mole moves. When it is running really fast,
you, too, have to run to keep pace.

Index